T0250926

Practical Manual on
Plant Cytogenetics

Practical Manual on Plant Cytogenetics

Ram J. Singh

CRC Press
Taylor & Francis Group
Boca Raton London New York

CRC Press is an imprint of the
Taylor & Francis Group, an **informa** business

CRC Press
Taylor & Francis Group
6000 Broken Sound Parkway NW, Suite 300
Boca Raton, FL 33487-2742

First issued in paperback 2021

ISBN-13: 978-1-03-209603-2 (pbk)
ISBN-13: 978-1-4987-4297-9 (hbk)

Library of Congress Cataloging-in-Publication Data

Names: Singh, Ram J., author.
Title: Practical manual on plant cytogenetics / Ram J. Singh.
Other titles: Plant cytogenetics
Description: Boca Raton : CRC Press, 2018. | Includes bibliographical references.
Identifiers: LCCN 2017036705 | ISBN 9781498742979 (hardback)
Subjects: LCSH: Plant cytogenetics--Laboratory manuals. | Plant genetics--Laboratory manuals.
Classification: LCC QK981.35 .S5625 2018 | DDC 572.8/2--dc23
LC record available at https://lccn.loc.gov/2017036705

Visit the Taylor & Francis Web site at
http://www.taylorandfrancis.com

and the CRC Press Web site at
http://www.crcpress.com

Dedication

To my wife, boys, and grandson (Kingston)

Contents

Foreword

Plant cytogenetics is one of the oldest disciplines of plant biology. Enormous number of techniques for studying plant chromosomes have been published and scattered throughout the literature. Ram J. Singh has undertaken the monumental task of assembling and systematically compiling these techniques into a comprehensive manual on plant cytogenetics. He has studied cytogenetics for almost 50 years and has made outstanding contributions. Furthermore, he has developed many of these techniques himself.

The manual begins with an elaboration of techniques to study mitotic and meiotic chromosomes including the so-called *squash* and *smear* techniques. This is followed by a discussion of the use of *fluorescence in situ* hybridization, which allows the study of structure of chromosomes. Next, the book addresses the technique of staining of pollens to determine whether they are fertile or sterile. We all know that cell division is critical for plant growth and reproduction and this chapter dealing with this topic is highly scholarly. The chapter on flow analysis and sorting of plant chromosomes is so thorough that it could have been published as a separate manual.

In my view, the chapter on modes of reproduction serves as an excellent reference source for the plant breeders. Karyotype analysis is fundamental to identifying individual chromosomes around the world; it is central to advances in cytogenetics. The chapter on genetic analysis and nature of gene action is admirable. Variations in chromosome structure and number provide tools for associating genes and linkage groups with respective chromosomes and chromosome arms is very thorough. We all know that most of the crop species have wild relatives that are useful resources for crop improvement. Plant breeders will benefit greatly from the chapter on wide hybridization.

I would like to congratulate Ram J. Singh for his love of labor in undertaking this monumental task of assembling and systematically collating the literature on advances in techniques of cytogenetics. This manual will serve as a standard refernce for cytogeneticists for years to come.

Gurdev S. Khush
University of California

Preface

A cell is regarded as the true biological atom.

George Henry Lewes
(The Physiology of Common Life, 1860, p. 297)

The cell is the life of all living organisms of Mother Earth. Robert Hooke (1665; http://www.gutenberg.org/files/15491/15491-h/15491-h.htm) coined the term "cell" based on the texture of cork. During the seventeenth to the nineteenth centuries, cytologists discovered staining chemicals, techniques, and instruments that could be used to identify chromosomes and to observe stages of mitosis, meiosis, and the life cycles of living organisms. Tremendous progress occurred in the discipline of cytology during the twentieth century, and several cellular organelles were identified. The cell consists of a nucleus, a nuclear membrane, a nucleolus, ribosomes, chromatin, a cell membrane, a cell wall, plasmodesmata, vacuoles, chloroplasts, mitochondria, leucoplasts, chromoplasts, Golgi bodies, cytoplasm, and a cytoskeleton. The functions of all the organelles are precisely controlled by mitosis and meiosis and regulated by the genes. The nucleus is the "heart" of a cell harboring the chromosomes that control and regulate the continuity of life.

The study of cells, and particularly of chromosomes, is called "cytology." Shortly after the rediscovery of Johann Gregor Mendel's (1885, 1886) laws of inheritance by Carl Correns, Hugo DeVries, and Erich von Tschermak in 1900, Walter Stanborough Sutton and Theodor Boveri (1903) independently reported the involvement of chromosomes in heredity. Wilson (1928) proposed the Sutton–Boveri theory of inheritance. Cytogenetics flourished once it was established that chromosomes carry genes. Thus, cytogenetics is a hybrid science that combines cytology (the study of chromosomes and other cell components) and genetics (the study of inheritance). Between 1910 and 1950, cytogeneticists discovered many genetic mechanisms using the higher organisms *Drosophila* (Thomas Hunt Morgan), *Datura* (Albert Francis Blakeslee), and maize (Barbara McClintock). These discoveries were used, initially, in conventional cytological techniques described in Chapter 2.

The chromosome is the vehicle and genes are the drivers controlling development of all living organisms. The normal route is controlled by reproduction, heredity, and development. Mitosis is responsible for the production of identical cells and meiosis is a key element for reproduction and heredity. Methodologies have been developed, refined, and modified to examine chromosomes, depending upon the organisms. At the same time, refinements in microscopes have allowed a better understanding of the structure and function of chromosomes.

Routine cytological techniques were developed that could be used for most organisms and protocols were modified whenever available techniques did not produce satisfactory results. Stains have been developed facilitating the identification of various structures of chromosomes, such as kinetochores, heterochromatin, euchromatin, nucleolar organizers, and chiasmata. Acetocarmine stains cytoplasm, the nuclear wall, nucleolus, and chromosomes, while the Feulgen stain is specific to the DNA (deoxyribose nucleic acid) of the chromosome. Between 1970 and 1980, Giemsa C- and N-banding techniques were developed, facilitating the identification of chromosomes at the mitotic metaphase in barley, rye, and wheat. Fluorescence *in situ* hybridization and genomic *in situ* hybridization are currently helping in identifying introgressed regions of donor chromosomes in recipient parents. The invention of flow cytometers has indeed enabled the sorting of chromosomes and the determination of DNA content of an organism.

Several classical books on cytology by world-renowned cytologists of the time, notably L. W. Sharp, C. D. Darlington, and A. K. Sharma, have been available for a long time. These books have been revised periodically as techniques improved. Further, many edited books on cytology are being published, but they lack basic background information published in the earlier, classic books. Finally, cytology has helped all of us in counting chromosomes of organisms, including those from higher plants, and many compendia, which include this information, have indeed been published.

This book is a unique laboratory manual of cytology; it begins with a brief introduction of the history of cytology and then flows into a description of classical methods (squash techniques) for handling plant chromosomes, smear techniques, fluorescence *in situ* hybridization, and genomic *in situ* hybridization. We briefly discuss pollen fertility, which is an important aspect of cytology, particularly for determining pollen stigma incompatibility. Cytological techniques modified for examining mitotic and meiotic stages are then described. Knowledge of sexual and asexual reproduction in higher plants and methodologies to determine the nature of plant sexual behavior are required and, thus, presented here. Karyotype analysis of chromosomes is relatively easy for plants with asymmetrical chromosomes compared to those with symmetrical chromosomes; chromosomes of several plants have been examined with a routine cytological method used by the author. The principle of locating genes on a particular chromosome by using aneuploidy is described. Wide hybridization technology using soybean and its tertiary gene pool species, *Glycine tomentella*, is diagrammatically shown step-by-step. We end this book with a glossary and with a brief description of terminology used in cytology.

This book is a laboratory manual intended for students and researchers interested in cytological techniques for handling plant chromosomes.

Ram J. Singh
University of Illinois

Acknowledgments

I was introduced to cytogenetics, particularly aneuploidy, during my MSc (Ag) final year by Professor Rishi Muni Singh, then of the Government Agricultural College, Kanpur, India. I became fascinated with cytogenetics, and fortunately, my fate took me to the barley cytogenetics laboratory of the late Professor Takumi Tsuchiya, a world-renowned barley cytogeneticist, at Colorado State University, Fort Collins, Colorado. I became a beneficiary of his expertise of aneuploidy in barley that broadened my knowledge and expertise, and coincidently, I became associated with Professor G. Röbbelen, University of Göttingen, Germany and Dr. G. S. Khush at the International Rice Research Institute, Las Baños, the Philippines.

I am profoundly grateful to C. de Carvalho, G. Chung, T. Endo, and B. Friebe for unselfishly providing their expertise to enrich Chapter 4. My special thanks go to J. Doležel for permitting the inclusion of information on flow karyotyping in Chapter 5. Credit goes to K. Kollipara for the diagrammatic sketches included in this manual.

I am extremely indebted to Govindjee, A. Lebeda, U. Lavania, and D. Walker for their constructive suggestions and comments that helped me greatly improve the clarity of the book. I am very grateful to various scientists and publishers who permitted me to use several figures and tables. However, I am solely responsible for any errors, misrepresentations, or omissions that may remain uncorrected in this book. I sincerely acknowledge the patience of my wife, Kalindi, who let me spend numerous weekends, nights, and holidays completing this *Practical Manual on Plant Cytogenetics*.

Author

Ram J. Singh, PhD, is a plant research geneticist at the USDA-Agricultural Research Service, Soybean/Maize Germplasm, Pathology, and Genetics Research Unit, Department of Crop Sciences, at the University of Illinois at Urbana-Champaign. Dr. Singh received his PhD in plant cytogenetics under the guidance of a world-renowned barley cytogeneticist, the late Professor Takumi Tsuchiya from Colorado State University, Fort Collins, Colorado. Dr. Singh isolated monotelotrisomics and acrotrisomics in barley, identified them by Giemsa C- and N-banding techniques, and determined chromosome arm-linkage group relationships.

Dr. Singh conceived, planned, and conducted quality pioneering research related to many cytogenetic problems in barley, rice, wheat, and soybean and has published in highly reputable national and international prestigious journals including: *American Journal of Botany, Caryologia, Chromosoma, Critical Review in Plant Sciences, Crop Science, Euphytica, Genetic Plant Resources and Crop Evolution, Genetics, Genome, International Journal of Plant Sciences, Journal of Heredity, Plant Molecular Biology, Plant Breeding,* and *Theoretical and Applied Genetics.* In addition, he has summarized his research results by writing eighteen book chapters. Dr. Singh has presented research findings as an invited speaker at national and international meetings. He has edited a series entitled Genetic Resources, Chromosome Engineering, and Crop Improvement. This series includes *Grain Legumes* (Volume 1), *Cereals* (Volume 2), *Vegetable Crops* (Volume 3), *Oil Seed Crops* (Volume 4), *Forage Crops* (Volume 5), and *Medicinal Plants* (Volume 6).

Dr. Singh has assigned genome symbols to the species of the genus *Glycine* and established genomic relationships among species by a cytogenetic approach. He has constructed a soybean chromosome map (for the first time) based on pachytene chromosome analysis and established all possible twenty primary trisomics and associated eleven of the twenty molecular linkage maps to the specific chromosomes. Dr. Singh has produced fertile plants with $2n = 40$, 41, and 42 chromosomes (for the first time) from an intersubgeneric cross of soybean cv. "Dwight" ($2n = 40$) and *Glycine tomentella*, PI 441001 ($2n = 78$). The screening of derived lines shows that useful genes of economic importance from *G. tomentella* have been introgressed into soybean. He holds U.S. patents on methods for producing fertile crosses between wild and domestic soybean species.

His book *Plant Cytogenetics* (first edition, 1993; second edition, 2003; third edition, 2017) is testimony to his degree of insight in research problems and creative thinking. His books are widely used worldwide by students, scientists, universities, industries, and international institutes. He is chief editor of the *International Journal of Applied Agricultural Research* and editor of *Plant Breeding and Biological Forum*—an international journal. He has won many honors and awards such as the Academic Professional Award for Excellence: Innovative & Creativity, 2000; College of ACES Professional Staff Award for Excellence—Research, 2009; The Illinois Soybean Association's Excellence in Soybean Research Award—2010; and Foreign Fellow, National Academy of Agricultural Sciences, India—2011.

1 Introduction

Cytology is a branch of biology that deals with the study of cells, chromosomes, and other cell organelles including their structure, function, and formation. Robert Hooke (1635–1703) coined the term *cell* in 1665 after observing a piece of cork under a microscope (Figure 1.1), and with keen observations, Hooke perceived the structure to be "all perforated and porous, much like a honeycomb, but pores were not regular; yet it was not unlike a honey-comb in these particulars (Figure 1.2)." He published his findings in his book *Micrographia* in 1665 (http://www.gutenberg.org/files/15491/15491-h/15491-h.htm#obsXVIII).

Antoni van Leeuwenhoek wrote a letter on June 12, 1716, in which he described his discoveries such as bacteria, free-living and parasitic microscopic protists, sperm cells, blood cells, and microscopic nematodes and rotifers (http://www.ucmp.berkeley.edu/history/leeuwenhoek.html). Robert Brown (December 21, 1773–June 10, 1858), a botanist and paleo botanist, described a cell nucleus and cytoplasmic streaming. He examined pollens from *Clarkia pulchella* suspended in water under a microscope and observed minute particles, now known as amyloplasts (starch organelles) and spherosomes (lipid organelles), ejected randomly from the pollen grains—known as Brownian motion (https://en.wikipedia.org/wiki/Robert_Brown_%28Scottish_botanist_from_Montrose%29).

Matthias Jakob Schleiden (April 5, 1804–June 23, 1881) examined plant structure and Theodor Schwann (December 7, 1810–January 11, 1882) inspected animal tissues under the microscope, and from their observation, they recognized the importance of cell nucleus and developed the cell theory. Theodor Schwann isolated an enzyme pepsin (https://en.wikipedia.org/wiki/Theodor_Schwann). Rudolf Ludwig Carl Virchow (October 13, 1821–September 5, 1902) was known as "the father of modern pathology" because his work discredited humourism, bringing more science to medicine. He linked the origin of cancer cells from normal cells. Robert Remak (July 26, 1815–August 29, 1865) showed the origins of cells from the division of preexisting cells.

Eduard Adolf Strasburger (February 1, 1844–18 May 18, 1912) was the first to describe accurately the embryo sac in gymnosperms (conifers) and angiosperms (flowering plants), and double fertilization in angiosperms. He conceived the modern law of plant cytology: "New cell nuclei can arise from the division from other nuclei" and coined the terms *cytoplasm* and *nucleoplasm*. Walther Flemming (April 21, 1843–August 4, 1905) and Édouard Joseph Louis Marie Van Beneden (March 5, 1846–April 28, 1910) independently investigated the process of cell division and the distribution of chromosomes to daughter nuclei, a process he called *mitosis* (from the Greek word for "thread") (https://archive.org/details/zellsubstanzker02flemgoog).

August Friedrich Leopold Weismann (January 17, 1834–November 5, 1914) developed germ plasm theory (i.e., that inheritance only takes place by means of the germ cells or the gametes [egg cells and sperm cells]). Based on this theory, he concluded that

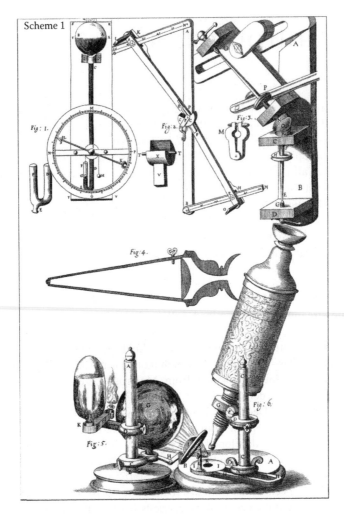

FIGURE 1.1 Hooke's microscope. (Available at http://www.gutenberg.org/files/15491/15491-h/images/scheme-01.png.)

two kinds of single reproductive cells are mutually attracted to one another, and then unite in the process of amphimixis, to constitute what we are accustomed to call the fertilized egg cell, which contains the combined hereditary substances of two individuals. He used the term *nuclear rod* or *idants* (chromosomes) as in "half the idants of two individuals become united in the process of amphimixis (sexual reproduction), and thus a fresh intermixture of individual characters results" (Weismann 1893). Heinrich Wilhelm Gottfried von Waldeyer-Hartz (6 October 1836–23 January 1921), a German anatomist, invented and coined the term chromosome (meaning colored bodies) in 1888 (https://en.wikipedia.org/wiki/Heinrich_Wilhelm_Gottfried_von_Waldeyer-Hartz).

Walther Flemming (April 21, 1843–August 4, 1905) used aniline stain to observe a structure which strongly absorbed basophilic dyes. In naming this, he coined the

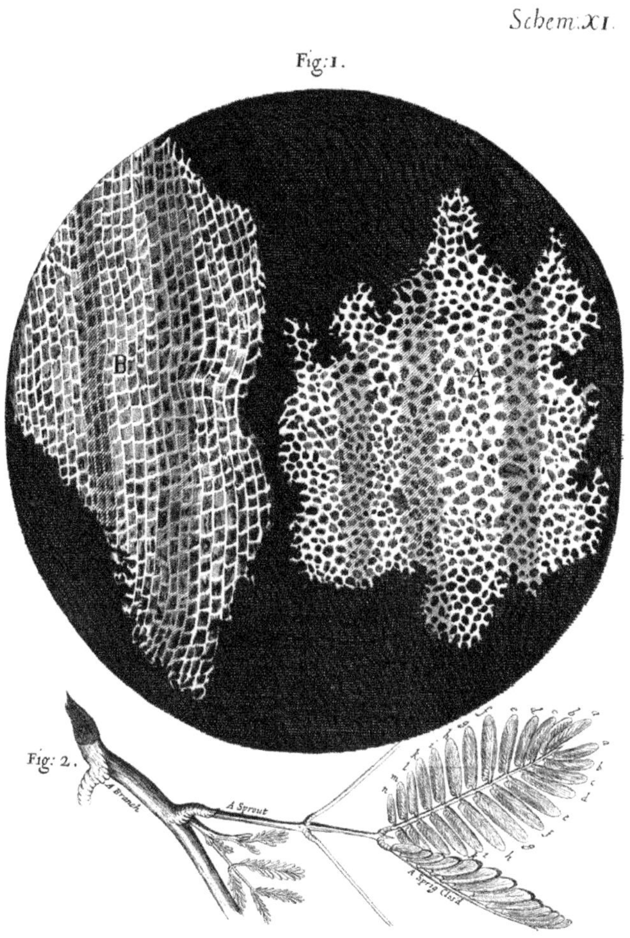

FIGURE 1.2 Hooke's observation on thin film of cork showing microscopical pores or cells (like wax in a honeycomb). (Available at http://www.gutenberg.org/files/15491/15491-h/images/scheme-11.png.)

term *chromatin* and was the first to conduct a systematic study of chromosomes during division. He called this process *mitosis* (Figure 1.3) and explained that the nucleus of reproductive cells divides twice, producing four genetically different daughter cells, each containing one set of chromosomes. However, he did not use the term meiosis (http://www.uni-kiel.de/grosse-forscher/index.php?nid=flemming&lang=e). Farmer and Moore (1905) described the terms *maiosis* or *maiotic* phase (reduction division) changed to *meiosis*.

Monumental progress in cytology was initiated shortly after the rediscovery of Mendel's laws of inheritance in 1900. Sutton (1903) reported the chromosomal basis of heredity (Sutton 1903). Boveri (1902) also published a paper. Edmund Beecher Wilson (1925) wrote a book entitled *The Cell in Development and Heredity*. On page 923, he wrote "Cytological Basis of the Mendelian Phenomena; The Sutton–Boveri Theory"

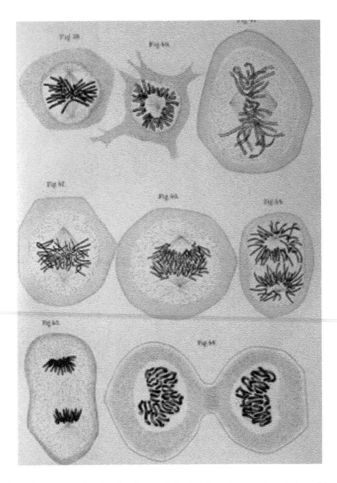

FIGURE 1.3 Diagrammatic sketch of mitosis by Walther Flemming. (From Flemming, W., Zellsubstanz, Kern, und Zelltheilung, 1882, F.C.W. Vogel, Leipzig.)

and gave credit to Sutton and Boveri for the Sutton–Boveri hypothesis. Since then, it has been known as the Sutton–Boveri hypothesis (also known as the chromosomal theory of inheritance or Sutton–Boveri theory). For some time, it was thought that Sutton and Boveri discovered the chromosomal theory of inheritance independently. Peters (1959) gave equal credit to Sutton and Boveri in his summary, with the statement that Sutton and Boveri published their findings "in the same year" but Lilian and Martins (1999) examined the facts of papers published between 1902 and 1903 and concluded that credit should be given only to Sutton, and not to Boveri, because Boveri did not publish any hypothesis of that kind during the relevant period.

Prior to development of the squash technique for studying chromosomes, published by Belling (1921), chromosome counts were conducted through staining of microtome sections, and sketches of cells and chromosomes were performed by Camera Lucida, known as Camera Lucida drawings, a technique which was patented in 1807 by William Hyde Wollaston (https://en.wikipedia.org/wiki/William_Hyde_Wollaston).

Numerous chemicals were tested for pretreatment and staining of the cells and chromosomes of plants, animals, and humans. Tjio and Levan (1956) determined the correct chromosome number ($2n = 56$) (Figure 1.4) of humans. Since then, human cytology has progressed at a much faster pace than plant cytology. The smear technique is routinely used for human chromosomes, while the squash technique is common for mitotic and meiotic plant chromosomes. Chromosome sizes and numbers in the plant kingdom are highly variable. Thus, cytological techniques are modified depending on the plant species. For example, ice-cold water pretreatment, acetic alcohol fixation (1:3), and acetocarmine/Feulgen staining produce good quality chromosome spreads for cereal (barley, wheat and rye, and oat) chromosomes, but for soybean chromosomes ($10 \times$ smaller than barley chromosomes: Figure 1.5)

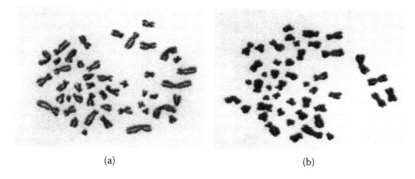

(a)   (b)

FIGURE 1.4 Colchicine-metaphase of human embryonic lung fibroblasts grown *in vitro*, (a) early metaphase and (b) full metaphase. (From Tjio, J.H., and Levan, A., *Hereditas*, 42, 1–6, 1956.)

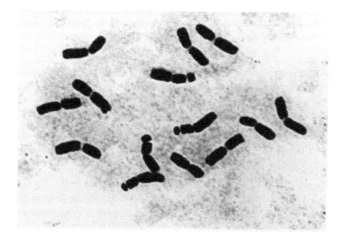

FIGURE 1.5 Acetocarmine stained barley chromosomes with $2n = 14$ after ice-cold water pretreatment.

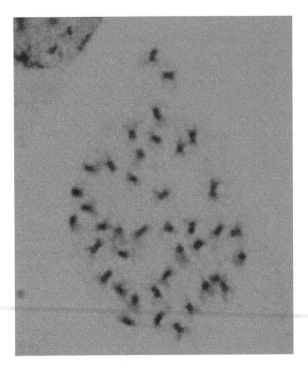

FIGURE 1.6   Carbol fuchsin-stained soybean chromosomes with $2n = 40$ after 8-hydroxyquinoline pretreatment.

pretreatment with 8-hydroxyquinoline at 13°C for 2–3 h, fixation in acetic alcohol (1:3) and Feulgen staining works the best for the soybean chromosomes (Figure 1.6). Since 1970, chromosome banding techniques, fluorescence *in situ* hybridization (FISH), genomic *in situ* hybridization, fiber FISH, flow karyotyping by flow cytometry, and chromosome image analysis are being used in plants. These techniques have been borrowed from human cytology (Singh 2017).

Cytology includes the handling of mitotic and meiotic chromosomes (pretreatment of tissues such as roots, fixation of roots and flower buds, and staining of roots and anthers), function and movement of chromosomes (mitosis and meiosis), determination of the number and structure of chromosomes, and karyotype analysis. The study of cytology is the foundation of genetics, and the combination of both, termed *cytogenetics*, give us a clear understanding of heredity, biosystematics, chromosomal aberrations, and the association of a gene with a particular chromosome using primary trisomics, genome analysis, and wide hybridization.

Cytological techniques developed rapidly following the publication of Belling's squash method. Chromosome cytology progressed rapidly in maize, *Datura*, faba bean, tomato, wheat, barley, and other plants. Several books on cytology have been published by world-renowned cytologists (Sharp 1921, 1943; Darlington and La Cour 1947, 1960, 1962, 1969, 1975; Sharma and Sharma 1965, 1972, 1980, 1994). Books on cytology have been edited by Adolph (1991), Fukui and Nakayama (1996), and Doležel et al. (2007). Simultaneously, cytologists published chromosome numbers of plants in books, journals, and newsletters (Appendix 1.1).

Many advances in cytological techniques using many important crops have been reported primarily in research papers. This book aims to compile all of the plant cytogenetic techniques previously published either in the books or in the research papers in one place in the form of a laboratory manual. The first part of the book describes standard cytological techniques routinely used by the students. The second part covers methods used for specific crops because one cytological method does not work for all plants. The third part covers cytogenetic techniques (cytology and genetics) for physically locating the genes within the specific chromosomes. At present, such a book is not available to students, teachers, or researchers.

REFERENCES

Adolph, K. W. (Ed.) 1991. *Advanced techniques in chromosome research.* (Ed.). Dekker, New York.

Belling, J. 1921. On counting chromosomes in pollen-mother cells. *The American Naturalist* 55(641): 573–574.

Boveri, Th. 1902. Ueber Mehrpolige Mitosen als Mittel zur Analyse des Zellkerns, Verb. d. Phys.–Med. Ges. zu Würzburg, N. F., Bd. XXXV.

Bretland Farmer, J., and J. E. S. Moore. 1905. On the meiotic phase (reduction divisions) in animals and plants: First division (heterotype) and second meiotic division (homotype). *Q. J. Microscopical Sci.* 192: 489–557.

Cave, M. S. 1958–1959. *Index to Plant Chromosome Numbers for 1956–1957 1 (1–2 and Supplement Prior to 1956).* California Botanical Society, Berkeley.

Cave, M. S. 1961–1965. *Index to Plant Chromosome Numbers for 1958–1964 1 (3–4), 2 (5–9).* University of North Carolina Press, Chapel Hill, NC.

Darlington, C. D., and E. K. Janaki Ammal. 1945. *Chromosome Atlas of Cultivated Plants.* Allen & Unwin, London.

Darlington, C. D., and L. F. La Cour. 1947. *The Handling of Chromosome.* 2nd ed. Allen & Unwin, London.

Darlington, C. D., and L. F. La Cour. 1960. *The Handling of Chromosome.* Macmillan, New York.

Darlington, C. D., and L. F. La Cour. 1962. *The Handling of Chromosome.* 4th ed. Allen & Unwin, London.

Darlington, C. D., and L. F. La Cour. 1969. *The Handling of Chromosome.* 5th ed, Allen & Unwin, London.

Darlington, C. D., and L. F. La Cour. 1975. *The Handling of Chromosome.* 6th ed., Wiley, New York.

Darlington, C. D., and A. P. Wylie. 1955. *Chromosome Atlas of Flowering Plants.* Allen & Unwin, London.

DeWolf, Jr. 1957. Chromosome Numbers in the Higher Plants. *Rhodora* 59: 241–244.

Dobešm C., and E. Vitek. 2000. *Documented Chromosome Number Checklist of Austrian Vascular Plants.*

Doležel, J., J. Greilhuber, and J. Suda, (Eds.) 2007. *Flow Cytometry with Plant Cells. Analysis of Genes, Chromosomes and Genomes.* Wiley-VCH Verlag GmbH & Co. KGaH, Weinheim.

Fay, M. 2011. Book review: Index to Plant Chromosome Numbers 2004–2006. *Bot. J. Linn. Soc.*, 226–227.

Fukui, K., and S. Nakayama, (Eds.). 1996. *Plant Chromosomes: Laboratory Methods.* CRC Press Inc. FL.

Goldblatt, P. 1981. Index to Plant Chromosome Numbers 1975–1978. *Monogr. Syst. Bot. Missouri Bot. Gard.* 6: 1–553.

Goldblatt, P. 1984. Index to Plant Chromosome Numbers 1979–1981. *Monogr. Syst. Bot. Missouri Bot. Gard.* 8: 1–427.

Goldblatt, P. 1985. Index to Plant Chromosome Numbers 1982–1983. *Monogr. Syst. Bot. Missouri Bot. Gard.* 13: 1–224.

Goldblatt, P. 1987. Index to Plant Chromosome Numbers 1984–1985. *Monogr. Syst. Bot. Missouri Bot. Gard.* 23: 1–264.

Goldblatt, P., and D. E. Johnson. 1990. Index to Plant Chromosome Numbers 1986–1987. *Monogr. Syst. Bot. Missouri Bot. Gard.* 30: 1–243.

Goldblatt, P., and D. E. Johnson. 1991. Index to Plant Chromosome Numbers 1988–1989. *Monogr. Syst. Bot. Missouri Bot. Gard.* 40: 1–238.

Goldblatt, P., and D. E. Johnson. 1994. Index to Plant Chromosome Numbers 1990–1991. *Monogr. Syst. Bot. Missouri Bot. Gard.* 51: 1–267.

Goldblatt, P., and D. E. Johnson. 1996. Index to Plant Chromosome Numbers 1992–1993. *Monogr. Syst. Bot. Missouri Bot. Gard.* 58: 1–276.

Goldblatt, P., and D. E. Johnson. 1998. Index to Plant Chromosome Numbers 1994–1995. *Monogr. Syst. Bot. Missouri Bot. Gard.* 69: 1–208.

Goldblatt, P., and D. E. Johnson. 2000. Index to Plant Chromosome Numbers 1996–1997. *Monogr. Syst. Bot. Missouri Bot. Gard.* 81: 1–188.

Goldblatt, P., and D. E. Johnson. 2003. Index to Plant Chromosome Numbers 1998–2000. *Monogr. Syst. Bot. Missouri Bot. Gard.* 94: 1–297.

Goldblatt, P., and D. E. Johnson. 2006. Index to Plant Chromosome Numbers 2001–2003. *Monogr. Syst. Bot. Missouri Bot. Gard.* 106: 1–242.

Goldblatt, P., and D. E. Johnson. 2010. Index to Plant Chromosome Numbers 2004–2006. *Regnum Veg.* 106: 1–242.

Goldblatt, P., and P. P. Lowry II. 2011. The *Index to Plant Chromosome Numbers* (Ipcn): Three Decades of Publication by the Missouri Botanical Garden Come to An End. *Ann Missouri Bot. Gard.* 98: 226–227.

Kalinka, A., G. Sramkó, O. Horváth, A. V. Molnár, and A. Popiela. 2015. Chromosome numbers of selected species of *Elatine* L. (Elatinaceae). *Acta Soc. Bot. Pol.* 84: 413–417.

Kihara, H., Y. Yamamoto, and S. Hosono. 1931. *A List of Chromosome –Numbers of Plants cultivated in Japan.* Yokendo, Japan.

Marhold, K., Mártonfi, P. M. Jun, and P. Mráz. 2007. *Chromosome number survey of the ferns and flowering plants of Slovakia.* Vydavateľstvo, Veda.

Moore, R. J. 1970. Index to Plant Chromosome Numbers for 1968. *Regnum Veg.* 68: 1–115.

Moore, R. J. 1971. Index to Plant Chromosome Numbers for 1969. *Regnum Veg.* 77: 1–116.

Moore, R. J. 1972. Index to Plant Chromosome Numbers for 1970. *Regnum Veg.* 84: 1–134.

Moore, R. J. 1973. Index to Plant Chromosome Numbers 1967–1971. *Regnum Veg.* 90: 1–539.

Moore, R. J. 1974. Index to Plant Chromosome Numbers for 1972. *Regnum Veg.* 91: 1–108.

Moore, R. J. 1977. Index to Plant Chromosome Numbers for 1973/74. *Regnum Veg.* 96: 1–257.

Packer J. G. 1964. Chromosome numbers and taxonomic notes on Western Canadian and Arctic plants. *Can. J. Bot.* 42: 473–494.

Peters, J. A. 1959. *Classic Papers in Genetics.* Prentice-Hall, Englewood Cliffs, NJ.

Rice, A., L. Glick, S. Abadi, M. Einhorn, N. M. Kopelman, A. Salman-Minkov, J. Mayzel, O. Chay, and I. Mayrose. 2015. *New. Phytologist.* 206: 19–26.

Sharma, A. K., and A. Sharma. 1965. *Chromosome Techniques: Theory and Practice.* Butterworths, London.

Sharma, A. K., and A. Sharma. 1972. *Chromosome Techniques: Theory and Practice.* 2nd ed. Butterworths, London.

Sharma, A. K., and A. Sharma. 1980. *Chromosome Techniques: Theory and practice.* 3rd ed. Butterworths, London.

Sharma, A. K., and A. Sharma. 1994. *Chromosome Techniques: A Manual.* Harwood Academic Publishers, Chur, Switzerland.

Sharp, L. W. 1921. *An Introduction to Cytology.* McGraw-Hill Book Company, Inc. New York.

Sharp, L. W. 1943. *Fundamentals of Cytology.* McGraw-Hill Book Company, Inc., New York and London.

Sutton, W. S. 1903. The chromosomes in heredity. *Biol. Bull.* 4: 231–250.

Tjio, J.H., and Levan, A., The chromosome number of man. *Hereditas*, 42: 1–6, 1956.

Weismann, A. 1893, *The Germ-Plasm: A Theory of Heredity.* Electronic Scholarly Publishing, prepared by Robert Robbins http://www.esp.org/books/weismann/germ-plasm/facsimile/

Wilson, E. B., 1925. Cell in Development and Heredity. *The Cell*, Macmillan, p. 1232.

2 Conventional Methods for Handling Plant Chromosomes

2.1 INTRODUCTION

Cytologists have devised cytological techniques from time to time to obtain precise information on chromosome numbers, chromosome structures, size and shape, and to examine the mechanism of cell division in plant species. These properties are studied by cytological techniques. The basic principles for handling the mitotic chromosomes of all plant species consist of the collection of specimens, pretreatment, fixation, and staining. For meiotic chromosomes, flower buds undergoing meiosis are collected in the morning and placed in a fixative, and anthers are stained in an appropriate solution. Slides for mitotic and meiotic chromosomes are prepared by a squash technique (Belling 1921). The cytological procedures are modified depending upon the crop species, the objective of the experiments, available facilities, and, above all, the personal preference of the cytologists (Sharma and Sharma 1965, 1972; Darlington and La Cour 1969; Fukui and Nakayama 1996; Jahier 1996; Singh 2017).

2.2 MITOTIC CHROMOSOMES

2.2.1 COLLECTION OF ROOTS

For cereal chromosomes, place the seeds on a moistened filter paper in a Petri dish. Keep the Petri dish in a dark cold room or refrigerator (0°C–4°C) for 5–7 days. The cold treatment facilitates uniform and rapid seed germination. Remove the Petri dish from the refrigerator and leave at room temperature (20°C–25°C) for germination. Collect roots that are 1–2 cm long. It should be kept in mind that roots should touch the filter paper; otherwise, the mitotic index will be very low (Figure 2.1a and b).

When it is desired to collect roots from plants growing in pots, care should be taken not to break the actively growing roots. After washing the soil from the roots, transfer roots to vials containing cold water. Keep the vials in ice-cold water in an ice chest.

For soybean and large-seeded legume chromosome analysis, seeds are geminated in a sand bench in the greenhouse; seedlings are ready to collect roots after about 1 week depending on the temperature of the room. Figure 2.2 shows 1-week-old soybean seedlings in a sand bench, and these are at the stage to collect the roots. Seedlings are carefully uprooted without damaging the actively growing cream-color roots (Figure 2.3a). Roots are healthy—the root tip is unbroken and is of cream color

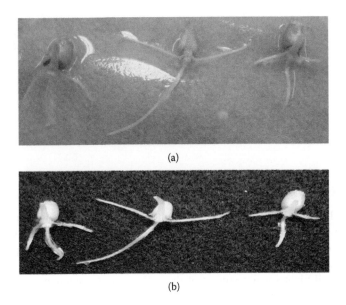

(a)

(b)

FIGURE 2.1 (a) Barley seeds germinating on a filter paper in a Petri plate and (b) wheat seeds germinated in a Petri plate (for photography, seedlings were moved to a black background).

FIGURE 2.2 Soybean seeds germinated in a sand bench in the greenhouse; seedlings are 5 days old and ready to collect the roots.

(Figure 2.3a; arrow). Transfer seedlings in a container with cold tap water (Figure 2.3b). Eppendorf tubes 1.5 mL are numbered in the laboratory and 1 mL ddH$_2$O is added. Root tips (three to five) are collected and transferred immediately in vials and the same number is assigned to the seedling label (Figure 2.4). Remove water from each tube and add 1 mL of 8-hydroxyquinoline and transfer the tubes to a heating block kept in a refrigerator at 13°C for 2–3 h. Generally, roots from cereals are collected in the laboratory after germinating the seeds in a Petri dish.

(a)   (b)

FIGURE 2.3   (a) Seedlings were uprooted carefully and gently to avoid breaking the actively growing roots (arrow) and (b) seedlings were transferred in a beaker filled with cold water.

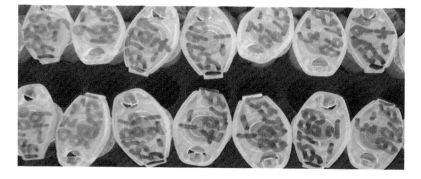

FIGURE 2.4   Three to five healthy roots were transferred in numbered 1.5 mL Eppendorf tubes that contained 1 mL distilled water.

2.2.2   Pretreatment of Roots

Pretreatment of roots is an essential step for studying the mitotic chromosomes. It performs several purposes: it stops the formation of spindles, increases the number of metaphase cells by arresting the chromosomes at the metaphase plate, contracts the chromosome length with distinct constrictions, and increases the viscosity of the cytoplasm. Numerous pretreatment agents, described below, have been developed.

2.2.2.1   Ice-Cold Water

Pretreatment with ice-cold water (0°C–2°C) is very effective and is widely used for cereal chromosomes. This particular pretreatment is preferred over other pretreatments when chromosomes of a large number of plants are studied. *This method of pretreatment is safe and should be tried first before heading to the chemicals.*

- Keep roots in numbered vials two-thirds filled with cold water.
- Transfer vials with roots into a container that will allow cold water to cover the top of the vials and further cover the vials with a thick layer of ice.
- Keep the container in a refrigerator for 12–24 h depending on the materials and experimental objectives. The recommended pretreatment period for barley and wheat chromosomes is 18–24 h. Longer pretreatment will shorten the chromosome length considerably.
- Chromosomes are usually not straight; kinetochores are not distinct and swollen. Mitosis is not completely arrested at metaphase, and few cells with anaphase and telophase are observed.

2.2.2.2   8-Hydroxyquinoline (C_9H_7NO)

Tjio and Levan (1950) were the first to recognize the usefulness of 8-hydroxyquinoline for chromosome analysis.

- Prepare aqueous solution by dissolving 0.5 g of 8-hydroxyquinoline in 1 L double distilled water (ddH_2O) at room temperature (RT) on cold stir plate, requires 24 h or more to dissolve.
- Make sure chemical is dissolved, filter in a colored bottle, and store in a refrigerator.
- Pretreat roots in 8-hydroxyquinoline for 2–3 h at 15°C–16°C. Warmer temperatures (RT) often yield a low frequency of metaphase cells in barley and also cause sticky chromosomes.
- Pretreatment with 8-hydroxyquinoline has been very effective for plants with small-size chromosomes like soybean. However, after staining, it makes primary (kinetochore) and secondary constrictions (nucleolus organizer region) and cytoplasm very clear.

2.2.2.3   Colchicine ($C_{22}H_{25}O_6N$)

- Colchicine is a toxic and poisonous natural product and secondary metabolite. It was originally extracted from plant *Colchicum autumnale* (meadow saffron).
  - It was originally used to treat gout.
  - Colchicine inhibits the formation of microtubule and is known as "mitotic poison."
- Colchicine prevents spindle formation, allows the chromosomes to be well spread in the cell, straightens the chromatids, allows the primary and secondary constrictions to become very noticeable, increases the number of metaphase chromosomes by preventing them from going to anaphase, and facilitates squashing and smearing. However, mitotic index is higher than ice-cold water and 8-hydroxyquiniline.

- Since colchicine in higher concentrations induces polyploidy, a low concentration (0.2%–0.5% for 1–3 h, at RT) is recommended for pretreatment.
- However, pretreatment of tissues at 8°C–9°C arrests higher frequency of metaphase cells than those at RT.
  - Store colchicine solution in a colored bottle in the refrigerator.
  - Treatment time of 1 h 30 min gives the best results for soybean chromosomes.
  - Roots should be washed thoroughly after colchicine pretreatment.
- This pretreatment facilitates better penetration of fixative at the subsequent stages of chromosome preparation.
- Colcemid (colchimid) is related to colchicine but it is less toxic. It also limits microtubule formation, inactivates spindle fiber formation, and arrests cells at metaphase. It is commonly used to arrest metaphase cells in mammals.

CAUTION: Since colchicine is a toxic chemical (carcinogenic and lethal poison), necessary precautions should be taken when it is used.

2.2.2.4 α-Bromonaphthalene ($C_{10}H_7Br$)

- The effect of α-bromonaphthalene is almost the same as that of colchicine.
- It is sparingly water soluble.
- A saturated aqueous solution is used in pretreatment for 2–4 h at RT (24°C ± 4°C).
- Prepare stock solution of 1 mL in 100 mL absolute ethanol.
- Store at RT; just before use, make a dilution of 10 μL stock in 10 mL distilled water.
- Treat roots for 18 h at 4°C.
- It is very effective for wheat, rye, barley, and other cereal chromosomes.

2.2.2.5 p-Dichlorobenzene ($C_6H_4Cl_2$)

- It is an organic compound and is colorless with strong odor.
- Like α-bromonaphthalene, it has a low solubility in water and is carcinogenic.
- It is very effective for plants with small-size chromosomes.
- Weigh 3 g of p-dichlorobenzene and add to it 200 mL distilled water.
- Incubate overnight at 60°C and then cool.
- Some crystals may remain undissolved. Shake thoroughly before using.
- Pretreat roots for 2–2 h 30 min at 15°C–20°C or at RT (Palmer and Heer 1973).
- p-Dichlorobenzene is effective for plants with both long and short chromosomes but period of pretreatment requires modification (Sharma and Sharma 1972).
- However, it is advantageous in case of plants with many and small chromosome complement.
- One of the serious drawback of this agent is that in the majority of cases, the period of treatment required is at least 3 h; temperature is also critical and cold treatment at 10°C–16°C is essential; and longer pretreatment produces chromosome fragmentation.

2.2.2.6   Nitrous Oxide (N₂O)

Nitrous oxide (laughing gas), a gaseous substance, was found to induce chromosome doubling and has spindle inactivation effect like colchicine (Östergren 1954).

Kato (1999) obtained the mitotic index 7.8% after pretreating maize roots with N_2O at 10 atm for 3 h, while the mitotic index in roots pretreated with 8-hydroxyquinoline (0.4%, 3 h) was 4.9%.

- Cut roots of maize from germinating seeds with a razor blade.
- Place on wet vermiculite in Petri dishes.
- Move Petri dishes into air-sealed container (10 cm inner diameter, 10 cm depth, with silicon rubber packing).
- Treat roots with pressurized N_2O gas (Figure 2.5).

Andres and Kuraparthy (2013) used the following protocol for pretreating the cotton root tips:

- Prepare 1.5 mL Eppendorf micro-centrifuge tubes prior to root tip excision.
- Make three to four holes on the top of each tube with a dissecting needle.
- Excise one to two actively growing root tips (2–4 cm long) and keep in the micro-centrifuge tubes and mist tubes with distilled water using a standard spray bottle.
- Place 1.5 mL tubes in a custom-built gas chamber.
- Apply nitrous oxide gas pressure of 160 psi (~10.9 atm) for 1 h 45 min.

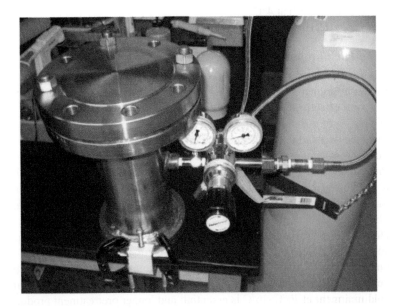

FIGURE 2.5   Nitrous oxide unit for pretreating the roots.

2.2.3 FIXATION

The science of chromosome study depends upon good fixative. The purpose of fixation is to fix or stop the cells at a desired stage of cell division without causing cell and chromosome distortion, swelling, or shrinkage of the chromosomes. Proper fixation should not be limited to the counting of chromosomes but it should increase the visibility of chromosomes, clarify the details of chromosome morphology such as primary and secondary constrictions, and differentiate euchromatic and heterochromatic regions.

A number of fixatives have been developed depending upon materials; monocotyledons and dicotyledons require different fixatives. Fixative chemicals have been classified into two categories:

1. The category one compounds cause pyknosis of chromosomes or detachment of nucleic acid from the protein thread.
2. The chemicals of the second category maintain the chromosome structure intact.

The primary prerequisite of a fixative is to study chromosomes in the possession of the property of precipitating chromatin. The fixative should provide rapid penetration into the tissues so that tissue is killed immediately and the cell divisions are stopped at their respective stages. Two types of fixative ingredients have been identified:

1. *Nonmetallic*: Ethyl alcohol, methyl alcohol, acetic acid, formaldehyde, propionic acid, picric acid, and chloroform
2. *Metallic*: Chromic acid, osmic acid, platinic chloride, mercuric chloride, uranium nitrate, and lanthanum acetate

The aim of a good fixative is to prevent aseptic roots. The most widely used fixatives have been described by Sharma and Sharma (1972) and are described below:

2.2.3.1 Carnoy's Solution I

1. Three parts ethanol (95%–100%)
2. One part glacial acetic acid

    Ethanol (CH$_3$CH$_2$OH): This is used as a major constituent in fixative used in various combinations and proportions with other acids. It has immediate penetration ability by dehydrating the tissues and also denatures the proteins. It can be used in combination with acetic acid, formaldehyde, and chloroform.

    Glacial acetic acid (CH$_3$CO$_2$H): Glacial acetic acid is routinely used in a fixing mixture in combination with ethanol and methanol. The term "glacial" is derived from the German term "ice-vinegar," which refers to the crystals that are created when acetic acid is cooled below the melting point; crystals are pure and contain no water. In combination with ethanol, it maintains the chromosome structure intact without distortion. Acetic acid is an excellent solvent of dyes such as carmine and orcein.

Note: This fixative is prepared fresh each time and is used for the fixation of roots. The material should be kept in the fixative for at least 24 h at RT and store at −20 °C.

2.2.3.2 Carnoy's Solution II

1. One part glacial acetic acid
2. Three parts chloroform
3. Six parts ethanol (95%–100%)

   Chloroform (CHCl₃): This is a toxic organic compound and colorless liquid. It is used in the above combination for fixing particularly wheat root tips. A modification of 1:3:4 has also been used for wheat chromosomes.

2.2.3.3 Propionic Acid Alcohol

1. One part propionic acid
2. Three parts ethanol (95%–100%)

   Propionic acid (CH₃CH₂COOH): Propionic is a clear liquid with a strong pungent and unpleasant smell and should be used under the fume hood. It is an excellent substitute for glacial acetic acid (*if one can tolerate the smell*). This fixative is very good for plants with small chromosomes. Roots are washed twice in 70% ethanol to remove the propionic acid. Materials can be stored long term at −20°C.

2.2.4   STAINING OF MITOTIC CHROMOSOMES

Chromosome number, morphology, structure, and behavior can be visualized only through a microscope after optimal staining of the chromosomes. A good quality staining agent specifically stains the chromosomes, differentiates euchromatin and heterochromatin, and provides clear cytoplasm with stained nucleoli (but not with the Feulgen stain). The following stains have been used extensively for staining plant chromosomes.

2.2.4.1 Preparation of Stain

1. Acetocarmine stain
   Ingredients
   For 1% stain:
   a. 1 g carmine powder (use carmine certified by the Biological Stain Commission)
   b. 45 mL glacial acetic acid
   c. 55 mL ddH₂O water
   Preparation of stain
   a. Heat to boil 100 mL 45% acetic acid under a fume hood.
   b. Add carmine powder to the boiling 45% acetic acid.
   c. Boil for 5–10 min, with occasional stirring until color becomes dark red.
   d. Cool and filter into a colored bottle and store in a refrigerator.

Note: Endo (2011) used a reflux condenser to prevent the solution from being boiled dry and boiled for 24 h; transfer to a bottle without filtration and store at RT.

The following URL (http://www.tcichemicals.com/eshop/en/us/commodity/ A0050/) provides ready-to-use acetocarmine solution, which comes in a colored 500 mL bottle. The concentration is 2% carmine, 10% acetic acid, and 88% water.

For preparing aceto-orcein, carmine is replaced by orcein in the same procedure.

2. Feulgen stain
   Ingredients
   a. 1 g basic fuchsin (use a stain commission type especially certified for use in the Feulgen technique)
   b. 200 mL ddH$_2$O
   c. 30 mL 1-N HCl
   d. 3 g potassium metabisulfite (K$_2$S$_2$O$_5$)
   e. 0.5 g activated charcoal
   Preparation of stain
   Method I:
   a. Dissolve 1 g basic fuchsin gradually in 200 mL of boiling ddH$_2$O and shake thoroughly.
   b. Cool to 50°C and filter.
   c. Add to filtrate 30 mL 1N HCl and then 3 g potassium metabisulfite (should not be from old stock).
   d. Close the mouth of the container with a stopper, seal with Parafilm, wrap container with aluminum foil, and store in a dark chamber at RT for 24 h.
   e. If the solution shows faint straw color, add 0.5 g activated charcoal powder, shake thoroughly, and keep overnight in a refrigerator (4°C).
   f. Filter and store the stain in a colored bottle in a refrigerator.
   Method II (J. Greilhuber, personal communication 2002):
   a. Dissolve 1 g basic fuchsin in 30 mL 1-N HCl at RT (3 h, stirring).
   b. Add 170 mL dH$_2$O + 4.45 g K$_2$S$_2$O$_5$ (3 h or overnight, stirring) and tighten vessel.
   c. Add 0.5 g activated charcoal (15 min, stirring) and filter by suction. The reagent is colorless. Avoid evaporation of SO$_2$ during all steps.

   Note: After discoloration, the solution has lost the red tint, but may be brown like cognac. These compounds are removed by the charcoal treatment. The use of aged K$_2$S$_2$O$_5$ can result in precipitation of white crystal needles (leuco-fuchsin), which are insoluble in SO$_2$ water but soluble in dH$_2$O. This can cause serious problems with washing out residual reagent from the sample. The SO$_2$ water washes (0.5 g K$_2$S$_2$O$_5$ in 100 mL dH$_2$O + 5 mL 1 M HCl) should always be done for 30–45 min for removing nonbound reagent from the stained tissue. This way a plasmatic background can be avoided.

   Note: Ready-to-use Feulgen stain is also available: Sigma, Schiff reagent, Fuchsin-sulfite reagent (lot # 51H5014).

3. Alcoholic hydrochloric acid–carmine stain
   Snow (1963) developed this stain based on the study of mitotic and meiotic chromosomes of several plant species.

Ingredients
a. 4 g (certified) carmine powder
b. 15 mL dH$_2$O
c. 1 mL concentrated HCl
d. 95 mL (85%) ethanol
Preparation of stain
a. Add 4 g certified carmine to 15 mL dH$_2$O in a small beaker.
b. Add 1 mL concentrated HCl. Mix well and boil gently for about 10 min with frequent stirring.
c. Cool, add 95 mL 85% ethanol, and filter.

4. Lacto-propionic–orcein stain
   Dyer (1963) used this particular stain for a large range of crops. This stain is useful for plants with small or numerous chromosomes. Chromosomes are intensely stained and cytoplasm remains clear.
Ingredients
a. 2 g orcein
b. 50 mL lactic acid
c. 50 mL propionic acid
Preparation of stain
a. Add 2 g natural orcein to 100 mL of a mixture of equal parts of lactic acid and propionic acid at RT.
b. Filter and dilute the stock solution to 45% with dH$_2$O.

5. Carbol-fuchsin stain
Ingredients-1
Solution A
a. 3 g basic fuchsin in 100 mL 70% ethanol
b. 10 mL 3% basic fuchsin
c. Add 90 mL of 5% phenol in distilled water

Note: Solution A can be stored in a refrigerator for a long time.

Solution B
a. 55 mL solution A
b. 6 mL glacial acetic acid
c. 3 mL formalin

Note: Solution B can be stored only for 2 weeks in a refrigerator.

Staining solution (100 mL)
a. 20 mL solution B
b. 80 mL 45% acetic acid
c. 1.8 g sorbitol
Ingredients-2
  This recipe has been devised to reduce the concentration of formalin. The higher concentration of formalin is toxic.

Solution A
a. 20 mL 3% basic fuchsin in 70% ethanol
b. 80 mL of 5% phenol in ddH_2O

Note: Solution A can be stored in a refrigerator for a long time.

Solution B
a. 55 mL solution A
b. 6 mL glacial acetic acid
c. 3 mL formalin

Note: Solution B can be stored only for 2 weeks in a refrigerator.

Staining solution (100 mL)
a. 10 mL solution B
b. 90 mL 45% acetic acid
c. 1.8 g sorbitol

CAUTION: Handle carbol fuchsin stain under a fume hood, store stain for at least 2 weeks at RT before use (aging of stain), and store working stain in a refrigerator.

6. Giemsa stain
Giemsa stain was discovered by a German chemist and bacteriologist, Gustav Giemsa; it is used in cytology. It is specific for the phosphate groups of DNA and attaches itself to regions of DNA where there are high amounts of adenine-thymine bonding (https://en.wikipedia.org/wiki/Giemsa_stain).
   Preparation of Giemsa stock solution from powder (Kimber et al. 1975)
a. 1 g Giemsa powder
b. 66 mL glycerin
c. 66 mL methanol
   Dissolve Giemsa powder in the glycerin at 60°C for 1 h with constant stirring. Add methanol and continue stirring at 60°C for 1 day (24 h). Filter and keep in a refrigerator. It can be kept for 1 or 2 months.

2.2.4.2 Staining of Roots

1. Acetocarmine staining
   a. Usually 1% acetocarmine stain has been very effective for somatic chromosomes of barley.
   b. Roots are transferred directly into the stain after pretreatment in ice-cold water. Roots are kept in the stain at RT for about 5–7 days for hardening the cells.
   c. This stain acts like a fixative and was routinely used for the analysis of F_1 and F_2 aneuploidy barley populations.
2. Feulgen staining
   a. Wash fixed root tips once with ddH_2O to remove the fixative for 15 min. If formaldehyde has been used in the fixative, wash roots in running tap water thoroughly for 1 h or wash three times for 15 min in ddH_2O.
   b. Hydrolyze roots in 1-N HCl (60°C) for 10 min. Instead of 1N HCl at 60°C, 5N HCl at 20°C for 60 min may be used. The latter is

recommended for quantitative Feulgen staining (J. Greilhuber, personal communication 2002).

c. After hydrolysis, rinse root tips in ddH$_2$O, remove excess water, and transfer root tips to the Feulgen stain.

d. Adequate staining can be achieved after 1 h at RT in the dark.

e. Cut stained region of root, place on a clean glass slide, add a drop of 1% acetocarmine or 1% propionic carmine, place a cover glass, and prepare slide by squash method.

Tuleen (1971) modified the above procedures for barley chromosomes. After pretreatment, transfer roots to 1-N HCl for approximately 7 min at 55°C. Transfer hydrolyzed roots immediately to Feulgen stain. Keep roots in stain at RT for 2 h. If it is desired to keep materials for a longer period, store in a refrigerator. Squash in 1% aceto carmine.

Palmer and Heer (1973) developed a cytological procedure by using Feulgen stain for soybean chromosomes after comparing several techniques. Their modified version is as follows:

1. Germinate seeds until roots are 7–10 cm long. Collect root tips after the first 3 h of the 30°C period. Seed germinated in a sand box or bench in a greenhouse yield a large number of secondary roots after 5–7 days of germination. It is advised to germinate seeds of soybean or large-seeded legumes in a sand bench or in vermiculite in the greenhouse.

2. Excise 1 cm of root tips, slit last 1/3 with a razor blade. Pretreat tips in covered vials in saturated solution PDB 15°C for 2 h. Pretreatment of roots with 0.1% colchicine for 1 h 30 min at RT proved to be better than PDB treatment. Pretreatment with 8-hydroxyquinoline at 16°C–18°C for 2–3 h has been found to be very efficient to arrest a large number of metaphase cells for soybean.

3. Wash root tips with ddH$_2$O and fix in freshly prepared 3:1 (95% ethanol:glacial acetic acid) for at least 24 h in covered vials at RT.

4. Wash root tips in ddH$_2$O, drain the water, and hydrolyze in 1N HCl 10–12 min at 60°C.

5. Wash root tips once in ddH$_2$O and place in Feulgen stain in covered vials for 1 h 30 min–3 h at RT.

6. Wash root tips in ice-cold ddH$_2$O.

7. Place root tips in pectinase in spot plates for 1–2 h at 40°C. After treatment in pectinase, root tips may be stored in 70% ethanol in covered vials at 4°C in a refrigerator, or slides may be made immediately.

   Note: This step may be omitted. After appropriate staining, wash roots with chilled distilled water once and store in cold water in a refrigerator. Chromosome staining is not distorted as long as roots are in chilled ddH$_2$O.

8. Place a root tip on a clean slide and remove root cap with a razor blade. Place less than 1 mm of root tip (dark purple region only) in a drop of propionic carmine or acetocarmine stain. Place a cover glass and prepare slide by squash method.

   For soybean, roots collected from sand bench grown 1-week-old seedlings are processed through Feulgen staining protocol, below, and tips are squashed in a drop of carbol fuchsin stain.

- Remove fixative (1 part glacial acetic acid + 3 parts 95% ethanol) from 1.5 mL Eppendorf tube.
- Add 1 mL ddH$_2$O and leave roots tips for about 10 min.
- Remove ddH$_2$O, add 1mL 1N HCl, place tube in a digital dry bath at 60°C for 10 min.
- Remove 1N HCl through pipette, add 1 mL ddH$_2$O, and leave the tube on a FlipStrip™ microtube rack for 15 min.
- Remove ddH$_2$O and add 1 mL Feulgen stain.
- Keep the root in the stain in the dark (e.g., in a drawer) for 1 h.
- Remove Feulgen stain, add 1 mL cold water (a wash bottle kept in a refrigerator), and store in a refrigerator.

2.2.5 Preparation of Chromosome Spread

2.2.5.1 Supplies

1. Microscope (Figure 2.6)
2. Glass slides (Figure 2.7)
3. Cover glass (Figure 2.8)
4. Forcep and needle (Figure 2.9)
5. Ethanol burner (Figure 2.10)
6. Oil immersion (Figure 2.11)

2.2.5.2 Chromosome Spreading

1. After Feulgen staining of roots, the actively growing tip stains bright red (Figure 2.12).
2. Remove root cap, squeeze out meristematic cells (darkly stained) or cut 1 mm tip, and keep tissues on the center of slide.
3. Add a drop of 45% acetic acid or carbol fuchsin stain.
4. Apply a cover glass in such a way that material should be in the center of the cover glass.
5. Heat on a low flame on an ethanol burner; be careful not to boil.
6. Tap gently on the specimen, reheat, lift cover glass from the top right gently to spread the cells, reheat, and tap once or twice.
7. Place a clean paper towel or filter paper, apply pressure from right hand thumb (if right handed) or left hand thumb (if left handed).

FIGURE 2.6 Olympus microscope for counting chromosomes; attached camera is on the top and is connected to the computer.

FIGURE 2.7   Glass slide of premium quality written on the box.

FIGURE 2.8   Premium cover glass; information is written on the box.

8. Make sure not to wiggle the thumb or move the cover glass.
9. This helps in removing excess stain.
10. Observe cells through the microscope first by 20× lens. Once a well-spread cell is located use 100× oil lens for detailed observation.
11. Take photograph(s) and keep the records.
12. Figure 1.5 shows a metaphase cell of barley stained by acetocarmine stain showing $2n = 14$ chromosomes.
13. Figure 1.6 shows a metaphase cell of soybean stained by Feulgen stain and root tip was squashed in carbol fuchsin stain.

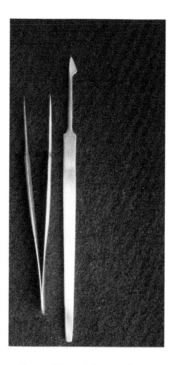

FIGURE 2.9 Forceps and needle used in picking up the roots and anthers, and preparation of chromosome spread.

FIGURE 2.10 Alcohol lamp to heat slide with the specimen and stain for spreading and staining of chromosomes.

FIGURE 2.11 Oil immersion used for the 100× oil lens.

FIGURE 2.12 Soybean roots treated with Feulgen stain. Note the dark purple stained root tip.

14. The above protocol was attempted for barley (Figure 2.13), wheat (Figure 2.14), faba bean (*Vicia faba*) (Figure 2.15), pea (Figures 2.16 and 2.17), pigeon pea (Figure 2.18), cowpea (Figure 2.19), lima bean (Figure 2.20), garlic (Figure 2.21), and shallot (Figure 2.22). Working with a wide range of crops suggests that the carbol fuchcin technique can be used on a routine basis for any crop plants.

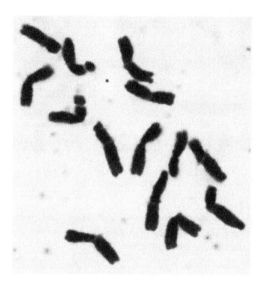

FIGURE 2.13 Mitotic metaphase chromosomes of barley stained with carbol fuchsin.

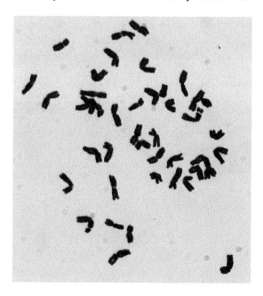

FIGURE 2.14 Mitotic metaphase chromosomes of hexaploid wheat with $2n = 42$ stained with carbol fuchsin.

FIGURE 2.15 Mitotic metaphase chromosomes of *Vicia faba* with $2n = 12$ chromosomes showing two longest chromosomes with satellite and ten acrocentric chromosomes. Chromosomes were stained with carbol fuchsin.

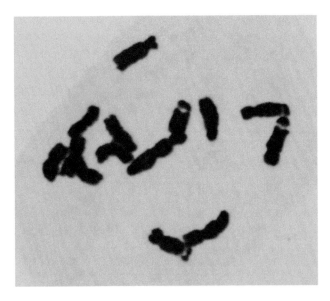

FIGURE 2.16 Mitotic metaphase chromosomes of pea with $2n = 14$ chromosomes showing two pairs of satellite chromosomes. Chromosomes were stained with carbol fuchsin.

FIGURE 2.17 Mitotic metaphase chromosomes of pea with 13 chromosomes, showing clearly two pairs of satellite chromosomes. Satellite is located on the long arm in both pairs.

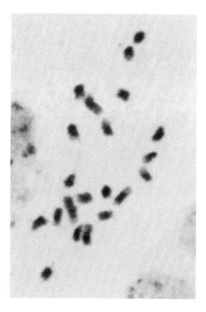

FIGURE 2.18 Mitotic metaphase chromosome of pigeonpea with $2n = 22$ chromosomes; chromosomes are symmetrical, and this is not a good material for karyotyping.

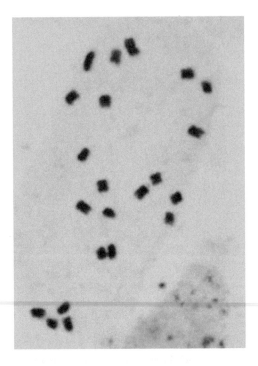

FIGURE 2.19   Mitotic metaphase chromosomes of cowpea with $2n = 22$ chromosome; like pigeonpea, all chromosomes are symmetrical, and this is not a good material for karyotyping.

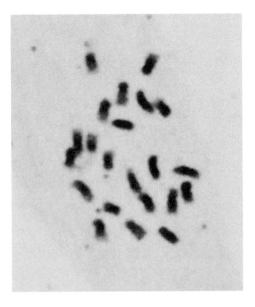

FIGURE 2.20   Mitotic metaphase chromosomes of lima bean with $2n = 22$ chromosomes; like pigeon and cowpea, all chromosomes are symmetrical, and this is not a good material for karyotyping.

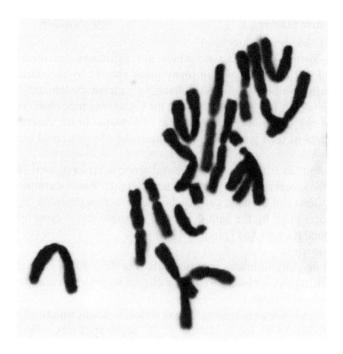

FIGURE 2.21 Mitotic metaphase chromosomes of scallion (green onion) with $2n = 16$. All chromosomes are clearly showing kinetochore, one pair of satellite chromosomes, mostly disassociated with the short arm.

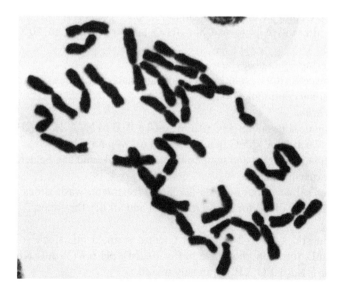

FIGURE 2.22 Mitotic metaphase chromosomes of shallot (a type of onion) with $2n = 32$ showing one pair of satellite chromosomes.

2.2.5.3   Giemsa Staining

1. Giemsa C-banding technique

The Giemsa C-banding technique has facilitated the identification of individual chromosomes in many plant species by its characteristic C- (constitutive heterochromatin) banding patterns (Singh and Tsuchiya 1981). Several minor modifications in the C-banding procedure have been made to obtain the maximum number of C-bands in the chromosomes. Procedures of seed germination and pretreatments described earlier can be followed.

a.   Fix roots in 3:1 (95% ethanol: glacial acetic acid) for at least 24 h.
b.   Soften cereal roots in 45% acetic acid or in 0.5% acetocarmine.
c.   Pectinase and cellulase can also be used to soften the roots.
d.   Prepare slide by the squash method. Remove cover glass by dry-ice method (Conger and Fairchild 1953).

Dehydration

i.   The majority of researchers treat slides in 95%–100% ethanol for 1 h (dehydration) to obtain C-bands in chromosomes *of Secale* species and *Hordeum vulgare.*
ii. After dehydration, keep air-dried slides at RT overnight (Singh and Röbbelen 1975). Linde-Laursen (1975) kept barley chromosome slides in a desiccator over silica gel for 1–2 weeks at 18°C.
iii. Process air-dried slides through the BSG (*Barium hydroxide/Saline/ Giemsa*) method.

Denaturation

i.   Prepare fresh saturated solution of barium hydroxide [5 g Ba (OH)$_2$. 8H$_2$O + 100 mL dH$_2$O]. Solution is filtered in a Coplin jar.
ii. Keep Coplin jar in a water bath (50°C–55°C) or at RT.
iii. An alternative is to use hot dH$_2$O and cool down to 50°C–55°C, or 20°C.
iv. Filtering is not necessary.
v.   Replace Ba (OH)$_2$ solution by cold water and rinse slides (J. Greilhuber, personal communication).

Renaturation

i.   Incubate slides in 2× SSC (0.3 M NaCl + 0.03 M Na$_3$C$_6$H5O$_7$.2H$_2$O), pH 7 to 7.6 at 60°C–65°C in a water bath or oven for 1 h.
ii. Incubation period and temperature are variable and can be determined by experimentation.
iii. After saline sodium citrate (2× SSC) treatment, wash slides in three changes of dH$_2$O for a total of 10 min and air dry the slides.

Staining

i.   Stain slides for 1–2 min with Giemsa stain, 3 mL stock solution + 60 mL Sörensen phosphate buffer (0.2M), pH 6.9 (30 mL KH$_2$PO$_4$ + 30 mL Na$_2$ HPO$_4$.2H$_2$O), freshly mixed.
ii. Monitor staining regularly.

iii. Leishman (*William Boog Leishman and Karl Reuter discovered independently in 1901*) and Wright (*James Homer Wright developed this stain in 1902*) stain also produces results similar to the Giemsa stain.

iv. After optimal staining, place slides quickly in dH_2O, air dry, store in xylene overnight, air dry again, and mount cover glass in euparal, Canada balsam, or permount.

v. Figure 2.23 shows mitotic metaphase chromosomes of barley after Giemsa C-banding staining.

vi. Giemsa C-banding technique has been used to identify rye chromosomes in wheat–rye disomic addition lines (Figure 2.24).

2. Giemsa N-banding technique

  This technique was originally developed to stain nucleolus organizing regions for mammalian and plant chromosomes.

  a. Incubate slides at $96 \pm 1°C$ for 15 min in 1N NaH_2PO_4 (pH 4.2 ± 0.2), adjust pH with 1N NaOH.

  b. Rinse thoroughly in distilled water and stain in Giemsa (dilute 1:25 in 1/15 M phosphate buffers, pH 7.0) for 20 min. Rinse slide in tap water and air dry.

    Gerlach (1977) modified the Giemsa N-banding technique for the staining of wheat chromosomes.

  a. Incubate air-dried slides in 1 M NaH_2PO_4 (pH 4.15) for 3 min at $94 \pm 1°C$

  b. Rinse slides in distilled water.

  c. Stain for 30 min with a solution of 10% Gurr® Giemsa R66 in 1/15 M Sörensen's phosphate buffer (pH 6.8).

  d. Rinse slides in tap water, mount in immersion oil.

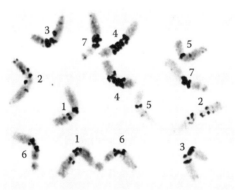

FIGURE 2.23 Giemsa C-banding of barley chromosomes ($2n = 14$) showing differentiated euchromatin and heterochromatin; all 14 chromosomes can be distinguished, which is not possible with roots processed through the Feulgen stain and after squashing cells in carbol fuchsin stain (Figure 2.13).

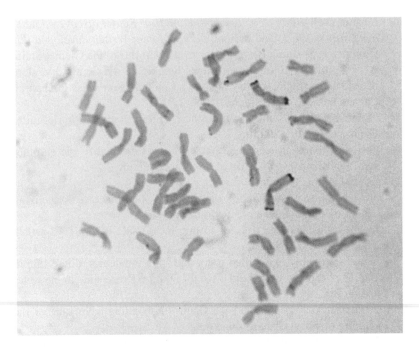

FIGURE 2.24 Giemsa C-banding of wheat–rye disomic addition line showing a pair of satellite rye chromosomes. Wheat chromosomes are not showing characteristic euchromatin and heterochromatin differentiation.

The Giemsa N-banding technique has been used with slight modifications:

Singh and Tsuchiya (1982b) suggested that the Giemsa banding technique should be considered as a qualitative tool to identify individual chromosomes, while conventional staining methods should be used as a quantitative approach to establish the standard karyotype. The combination of acetocarmine and Giemsa staining demonstrated that karyotype analysis could be conducted with greater precision than was previously possible. The kinetochore expressed more diamond-like structures in barley with the N-banding technique (Figure 2.25) than those observed by C-banding (Figure 2.23).

  i. Collect barley roots, pretreat with ice-cold water, and fix according to the procedures described earlier.

  ii. Transfer fixed roots to acetocarmine (0.3%) for about 2–3 h.

  iii. Prepare slides by the squash method.

  iv. Photograph the cells with well-spread chromosomes by phase contrast lens.

  v. Remove cover glass by the dry-ice method.

  vi. Place slides in 96% ethanol for 2–4 h. Endo and Gill (1984) treated wheat chromosome slides with hot (55°C–60°C) 45% acetic acid for 10–15 min.

  vii. Air dry slides overnight at RT. Linde-Laursen (1975) placed slides in a desiccator over silica gel for 2–4 weeks.

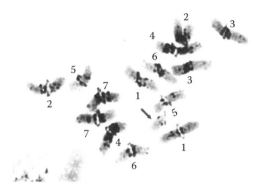

FIGURE 2.25 Giemsa N-banding of barley chromosomes with $2n = 14 + 1$ telocentric chromosome (arrow). N-banding procedure expresses a diamond shape centromere structure while, C-banding procedure does not (Figure 2.23).

    viii. Incubate slides in 1 M NaH_2Po_4, pH 4.15 for 1.5–2 min (5 min for barley) at $94 \pm 1°C$.

      ix. Rinse slides briefly in dH_2O, air dry.

       x. Stain for 20 min to 24 h in 1% Giemsa (Sigma CAS Number 51811-82-6) in 1/15 M Sörensen's phosphate buffer.

     xi. After optimal staining of the chromosomes, which were previously photographed from the acetocarmine preparation, rinse slides in dH_2O, air dry, keep in xylene overnight, air dry again, and mount in Permount.

    xii. Endo (2011) modified the Giemsa N-banding technique described in this chapter. However, he observed similar banding patterns for barley chromosomes to those reported by Singh and Tsuchiya (1982a).

3. HCl–KOH–Giemsa (HKG) technique

    Shang et al. (1988) claimed that the HKG technique (HCl–KOH–Giemsa) produces well separated and sharply banded chromosomes, including centromeric bands of wheat chromosomes and that the results are more highly reproducible than C- or N-banding techniques.

    a. Pretreat roots for 2 h 30 min with an aqueous solution of α-bromonaphthalene (0.01 mL stock solution in 10 mL water; stock solution: 1 mL α-bromonaphthalene in 100 mL absolute ethanol).

    b. Wash roots twice in distilled water and hydrolyze with 5N HCl for 20 min at RT.

    c. Wash roots twice in distilled water, store material in 45% acetic acid.

    d. Clean slides in 95% ethanol and wipe dry.

    e. Prepare slide by taking 1 mm meristematic part of the root in 45% acetic acid.

    f. Remove cover glass by liquid nitrogen. Cover glass can be removed by placing prepared slides in liquid nitrogen, by dry ice or by placing in a

−80°C freezer. The purpose is not to lose the specimen or to distort the chromosome spread.

g. Air dry slide and store for 3–7 days.

h. Treat air-dried slide with 1N HCl at 60°C for 6 min and wash slides four times in distilled water for a total of 10 min at RT.

i. Air dry slide for half a day, dip slide into fresh 0.07 N KOH for 20–25 s followed by dipping into 1/15 M Sörensen's phosphate buffer (pH 6.8) for 5–10 s with shaking.

j. Stain in 3% Gurr improved Giemsa stain solution (3 mL stain in 100 mL 1/15 M Sörensen's phosphate buffer, pH 6.8) for 1–2 h or until proper staining is reached. Keep slides in the stain for 1–2 days without over staining.

k. Rinse slide in distilled water, air dry and mount in a synthetic resin (Preserveaslide, Mathesin, Coleman and Bell Co.).

4. Modified HKG-banding technique

The following explains the modified HKG-banding technique (Carvalho and Saraiva 1993):

a. Hydrolyze 1–5-day-old slides in 1-N HCl at 60°C for 4–6 min.

b. Wash slides four times in dH_2O for a total of 10 min.

c. Immerse slides briefly in 0.9% NaCl, plunge 10 times in 70% ethanol, dry slides on a hot plate (surface temperature 50°C) for few seconds.

d. Immerse slides in 0.06N KOH solution for 8–12 s with continuous agitation at RT.

e. Wash slides in 70% ethanol (two changes) and 100% ethanol and transfer to methanol: acetic (8:1).

f. Air dry slides on a hot plate for few minutes, stain according to procedure described for the modified C-banding procedure, and observe through microscope (Figure 2.26).

5. Giemsa G-banding technique

The G-banding technique was originally developed to identify human chromosomes. Kakeda et al. (1990) perfected the G-banding method for maize chromosomes.

a. Germinate maize seeds in small pots filled with moist vermiculite for 2 days at 32°C under continuous light.

b. Excise three root tips about 1 cm long from each seed and pretreat either with a 0.05% colchicine solution or 0.05% colchicine solution containing either 10 ppm actinomycin D or ethidium bromide for 2 h at 25°C.

c. Dip root tips in the Ohnuki's hypotonic solution (55 mM KCl, 55 mM $NaNO_3$, 55 mM CH_3COONa, 10:5:2) for 30 min to 1 h at 25°C.

d. Fix root tips with methanol: acetic (3:1) for 1–4 days in a freezer (−20°C) or for at least 2 h at 25°C to actinomycin D pretreated ones.

e. Remove meristem cells with a tweezer in a drop of fresh fixative or macerate enzymatically.

Note: For enzyme maceration, wash fixed root tips for about 10 min and macerate in an enzyme mixture (2% Cellulase RS [Yakult Honsa Co., Ltd, Tokyo]

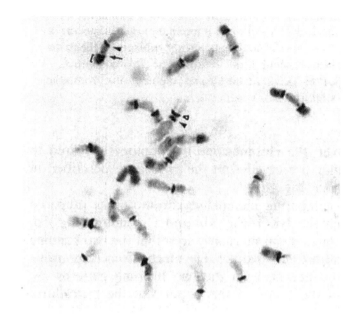

FIGURE 2.26 Modified Giemsa HKG technique for identifying maize chromosomes. (From Carvalho, C.R.D., and L.S. Saraiva, *Heredity,* 70, 515–519, 1993.)

and 2% MacerozymeR-200 [Yakult Honsa, Co., Ltd, Tokyo], pH 4.2) for 20–60 min at 37°C in a 1.5 mL Eppendorf tube. Rinse root tips with dH₂O two or three times. Pick up macerated roots tips with the help of a Pasteur pipette and place it on a glass slide.

f. Cut actively growing tip into small pieces with sharp-pointed tweezers with the addition of fresh fixative.

g. Observe slide under a phase contrast microscope, select slide with well-spread chromosomes and air dry for about 2 days in an incubator at 37°C.

h. Stain samples, prepared by the actinomycin D, directly in 10% Wright solution diluted with 1/15 M phosphate buffer (pH 6.8) for 10 min at 25°C, wash, and air dry slides.

i. For slides prepared by enzyme maceration, fix sample again in a 2% glutaraldehyde solution diluted with phosphate buffer for 10 min at 25°C and wash.

j. Immerse postfixed slides either in 2% trypsin (MERCK, Art. 8367) dissolved in PBS (pH 7.2) for 10 min at 25°C or in 0.02% SDS dissolved TRIS-HCl buffer (20 mM, pH 8.0) for 2–25 min at 25°C.

k. Wash slides briefly and air dry.

l. Stain slides in 5% Wright solution in 1/30M phosphate buffer (pH 6.8) for 5 min.

m. Observe G-banding pattern through a microscope (Figure 2.27).

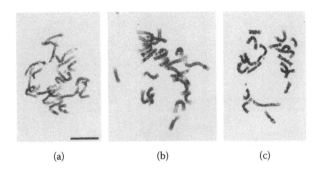

(a)   (b)   (c)

FIGURE 2.27   (a–c) Giemsa G. banding pattern of maize chromosomes. (From Kakeda et al. *Theor. Appl. Genet.*, 80, 265–270, 1990.)

2.3   MEIOTIC CHROMOSOMES

2.3.1   COLLECTION OF FLOWER BUDS

Meiotic stages are examined in pollen mother cells (PMCs) in flower buds undergoing meiosis during the morning (9 a.m. to 10 a.m.). It is extremely difficult to predict the precise meiotic stages in the reproductive organs of the plants because it varies from crop to crop, the time of meiosis initiation, temperature, light, growing conditions, and health of plants. Meiotic chromosome pairing is conducted primarily during prophase I preferentially at diakinesis, metaphase-I, and anaphase-I. However, chromosome pairing at pachynema cannot be studied in all the crops. However, it has been successfully accomplished in maize, tomato, rice, and soybean.

- In cereals, spikelets undergoing meiosis are collected between 8 a.m. and 9 a.m.
- Figure 2.28 shows the stage of spikes for wheat undergoing meiosis.
- Remove leaf sheath and examine the meiotic stages (Figure 2.29).
- It is suggested to start from the bottom of the spike and move up.
- Attempt a similar approach to oats (Figures 2.30 and 2.31).
- One anther is used to check the meiotic stage.
- If a desired stage is detected, remaining anthers are placed in a fixative.
- Preselection of anthers helps in identifying the precise meiotic stage for collecting the spikelets and panicles in cereals, and flower buds in legumes or inflorescence.
- It is not an easy to catch meiotic stages in soybean because the flower buds (two to five) are axillary and undergo meiosis when very young.
- Anthers are arranged in two whorls; the outer five anthers are older than the four inner anthers (Figure 2.32); and one anther is always under the stigma. Thus, meiotic stages are more advanced in the anthers of the outer whorl than those in the inner whorl.

FIGURE 2.28 Stage of wheat spikelets for examining meiotic chromosomes.

FIGURE 2.29 Three spikelets of wheat ready to submerge in the fixative.

FIGURE 2.30 Stage of four panicles of oats for examining meiotic chromosomes.

FIGURE 2.31 Three panicles of oats ready to submerge in the fixative, remove the top older region of panicle.

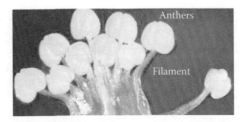

FIGURE 2.32 Anthers of soybean attached with the filament. These anthers have passed the meiotic stages and are ready to produce mature pollen, note the arrangement of anthers in two whorls (five outer side and four between under the first whorl. One anther (right) is always under the stigma).

2.3.2 Fixation

Preselection of anthers guides us to fix panicles/spikelets/inflorescence for studying the meiotic stages. The most widely used fixatives, like root tips, are as follows:

2.3.2.1 Carnoy's Solution 1

1. Three parts ethanol (95%–100%)
2. One part glacial acetic acid

This fixative is mixed prior going to the greenhouse to collect the specimens. This fixative is the foundation, and other fixatives evolved, depending on the materials, did not produce satisfactory results. Specimens are transferred to 70% ethanol for future use and stored in a refrigerator. This fixative for barley produced satisfactory results after staining anthers in acetocarmine.

2.3.2.2 Carnoy's Solution 2

1. One part glacial acetic acid
2. Three parts chloroform
3. Six parts ethanol (95%–100%)

This fixative is customarily used by wheat cytologists. A modification of 1:3:4 is widely used without any difference in the results. Wash spike twice in 70% ethanol to remove chloroform because this chemical is toxic. Store fixed specimen in 70% ethanol in a refrigerator.

2.3.2.3 Propionic Acid Alcohol

1. One part propionic acid
2. Three parts ethanol (95%–100%)
3. 1 g/100 mL fixative ferric chloride ($FeCl_3$)

This fixative is great for plants with small chromosomes like soybean (1.42–2.84 μm). Fixative should be prepared under the fume hood because propionic acid has a pungent and unpleasant smell. Fixative color turns yellow–orange. Fix inflorescence for over 24 h at RT. Wash twice with 70% ethanol under the fume hood and store in 70% ethanol in a refrigerator.

2.3.3 STAINING OF MEIOTIC CHROMOSOMES

- Screen from one anther from fixed panicle/spike/inflorescence specimen for desired meiotic stages such as pachynema/diakinesis/metaphase –I/anaphase-I.
- Transfer remaining anthers either in 1% acetocarmine or in 1% propiono carmine. Propiono carmine is used for plants with small chromosomes such as rice and soybean. Cytoplasm is clear with propiono carmine, and the pachynema chromosome is clearly differentiated by euchromatin and heterochromatin.
- Keep anthers in the stain for 1 week in a refrigerator.

2.3.4 PREPARATION OF CHROMOSOME SLIDE

- Remove one anther from the stain, place on a clean glass slide, add a drop of 45% acetic acid, and heat on a low flame, but make sure not to boil.
- Place 18 mm × 18 mm × 1 mm cover glass and make sure anther is in the center of the cover glass.
- Reheat on the flame and again, make sure not to boil.
- Tap once or twice on the anther to release the PMCs outside the anther.
- Apply pressure on the specimen and be careful not to move the cover glass.

2.3.5 OBSERVATION OF CELLS

- Observe slide through the microscope. It is recommended to screen slide at the periphery of the specimen.
- Screen cells by 20×/0.75× lens and once a cell with well-spread chromosomes is identified, observe the cell by 100× oil immersion lens.
- The optimum stain of diakinesis in barley (Figure 2.33), and pachynema chromosomes of rice (Figure 2.34) and soybean (Figure 2.35) are achieved without stained cytoplasm showing euchromatin and heterochromatin differentiation in pachynema chromosomes.

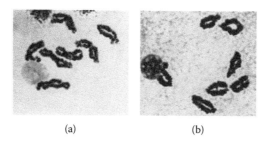

(a)   (b)

FIGURE 2.33 Meiosis in barley. Cell in (a) shows early stage of diakinesis with two bivalents associated with the nucleolus; (b) a cell at the end of diakinesis shows one pair associated with the nucleolus and six bivalents show terminalized chiasma.

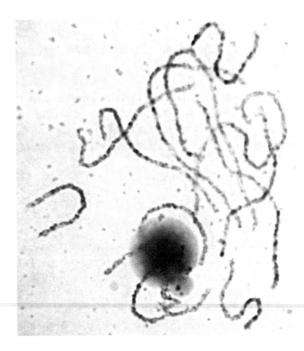

FIGURE 2.34   Pachynema chromosomes of rice; few chromosomes are clearly differentiated with euchromatin and heterochromatin.

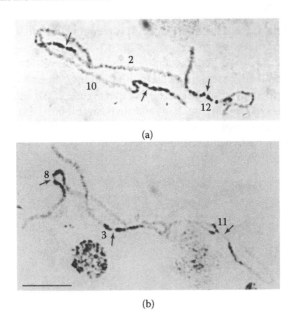

FIGURE 2.35   (a) Pachynema chromosomes of soybean showing kinetochores flanked by heterochromatin, and (b) chromosomes have been identified by numbers based on karyotype analysis.

2.4 SUPPLIES

- Microscope (Figure 2.6)
- Slide glass (Figure 2.7)
- Cover glass (Figure 2.8)
- Forceps and needle (Figure 2.9)
- Ethanol lamp (Figure 2.10)
- Oil immersion (Figure 2.11)
- Lens cleaner paper (Olympus Optical Co. Ltd.; CL-Tissue [M97]; AX6476; C-0100 030007)
- Ethanol (100%)
- Glacial acetic acid
- Propionic acid
- Ferric chloride ($FeCl_3$)
- 8-hydroxyquinoline
- Glass beakers (1000, 500, 250, 100, 50, 25, and 10 mL)
- Measuring glass cylinder (1000, 500, 250, 100, 50, 25, and 10 mL)
- Carmine
- Chloroform
- Giemsa (Sigma CAS Number 51811-82-6)
- Potassium metabisulphite ($K_2S_2O_5$)
- Activated charcoal
- Hydrochloric acid (HCl)
- Whatman filter paper
- Parafilm
- Stir heating block
- Fuchsin-sulfite reagent (Sigma lot # 51H5014)
- Lactic acid
- Phenol
- Formalin
- Sorbitol
- Petri plate
- Razor blade
- Pectinase
- Dry ice
- Barium hydroxide [Ba $(OH)_2$ $8H_2O$]
- Sodium chloride (NaCl)
- Sodium citrate dihydrate ($Na_3C_6H5O_7.2H_2O$)
- Water bath
- Oven
- Sörensen phosphate buffer
  (Potassium phosphate monobasic, KH_2PO_4;
  sodium phosphate dibasic dihydrate, $Na_2 HPO_4.2H_2O$)
- Sodium phosphate monobasic (NaH_2PO_4)
- Euparal, Canada balsam, or Permount
- Giemsa Gurr stain solution R66 (Fisher Scientific Catlog No. 50-300-415)

- Giemsa (Sigma CAS Number 51811-82-6)
- Potassium hydroxide (KOH)
- Actinomycin D
- Ethidium bromide
- Ohnuki hypotonic solution
  55 mM KCl
  55 mM $NaNO_3$
  55 mM CH_3COONa
  (10:5:2)
- Methanol
- Cellulase RS (Yakult Honsa Co., Ltd, Tokyo)
- Macerozyme R-200 (Yakult Honsa Co., Ltd, Tokyo)
- Eppendorf tube 1.5 mL
- Pasteur pipette
- Tweezer (sharp-pointed)
- Glutaraldehyde
- Trypsin (MERCK, Art. 8367)
- Phosphate buffer saline (PBS)
  For 1 L of 1× PBS, prepare as follows:
  - Start with 800 mL of distilled water:
  - Add 8 g NaCl
  - Add 0.2 g KCl
  - Add 1.44 g Na_2HPO_4
  - Add 0.24 g KH_2PO_4
  - Adjust the pH to 7.4 with HCl
  - Add distilled water to a total volume of 1 L

REFERENCES

Andres, R. J., and V. Kuraparthy. 2013. Development of an improved method for mitotic metaphase chromosome preparation compatible for fluorescence in situ hybridization in cotton. *Jour. Cotton Sci.* 17: 149–156.

Belling, J. 1921. On counting chromosomes in pollen – mother cells. *Amer. Nat.* 55: 573–574.

Carvalho, C. R. D., and L. S. Saraiva. 1993. A new heterochromatin banding pattern revealed by modified HKG banding technique in maize chromosomes. *Heredity* 70: 515–519.

Conger, A. G., and L. M. Fairchild. 1953. A quick-freeze method for making smear slides permanent. *Stain Tech.* 28: 281–283.

Darlington, C. D., and L. F. La Cour. 1969. *The Handling of Chromosomes*, 5th ed. George Allen & Unwin Ltd., London.

Dyer, A. F. 1963. The use of lacto-propionic orcein in rapid quash method for chromosome preparations. *Stain Tech.* 38: 85–90.

Endo, T. R., and B. S. Gill. 1984. Somatic karyotype, heterochromatin distribution and nature of chromosome differentiation in common wheat, *Triticum aestivum* L. em Thell. *Chromosoma (Berlin)* 89: 361–369.

Endo, T. R. 2011. Cytological dissection of the Triticeae chromosomes by the gameto-cidal system. *In Plant Chromosome Engineering: Methods and Protocols, Methods in Molecular Biology*, J.A. Birchler (Ed.), Springer Science + Business Media, LLC, 701: 247–257.

Fukui, K., and S. Nakayama. 1996. *Plant Chromosomes: Laboratory Manual*. K. Fukui and S. Nakayama (Eds.), CRC Press, Boca Raton, FL.

Gerlach, W. L. 1977. N-banded karyotypes of wheat species. *Chromosoma (Berlin)* 62: 49–56.

Jahier, J. 1996. *Techniques of Plant Cytogenetics*. (Ed. J. Jahier). Science Publishers, Inc. New Hampshire, Lebanon.

Kakeda, K., H. Yamagata, K. Fukui, M. Ohno, K. Fukui, Z, Z. Wei, and F. S. Zhu. 1990. High resolution bands in maize chromosomes by G-banding methods. *Theor. Appl. Genet.* 80: 265–272.

Kato, A. 1999. Air drying method using nitrous oxide for chromosome counting in maize. *Biotech. Histochem.* 74: 160–166.

Kimber, G., B. S. Gill, J. M. Rubenstein, and G. L. Barnhill. 1975. The technique of Giemsa staining of cereal chromosomes. *Res. Bull.* 1012: 3–6.

Linde-Laursen, I. B. 1975. Giemsa C-banding of the chromosomes of 'Emir' barley. *Hereditas* 81: 285–289.

Östergren, G. 1954. Polyploids and aneuploids of Crepis capillaris produced by treatment with nitrous oxide. *Genetica*, XXVII: 54–64.

Palmer, R. G., and H. Heer. 1973. A root tip squash technique for soybean chromosomes. *Crop Sci.* 13: 389–391.

Shang, X. M., R. C. Jackson, and H. T. Nguyen. 1988. Heterochromatin diversity and chromosome morphology in wheats analyzed by the HKG banding technique. *Genome* 30: 956–965.

Sharma, A. K., and A. Sharma. 1965. *Chromosome techniques - Theory and Practice*. Butterworths: London.

Sharma, A. K., and A. Sharma. 1972. *Chromosome techniques Theory and Practice*. Butterworths & Co (Publishers) Ltd, London.

Snow, R. 1963. Alcoholic hydrochloric acid-carmine as a stain for chromosomes in squash preparations. *Stain Tech.* 38: 9–13.

Singh, R. J. 2017. *Plant Cytogenetics*. 3rd ed. CRC Press (Taylor & Francis Group) Boca Raton, FL.

Singh, R. J., and G. Röbbelen. 1975. Comparison of somatic Giemsa banding pattern in several species of rye. *Z. Pflanzenzüchtg.* 75: 270–285.

Singh, R. J., and T. Tsuchiya. 1981. Identification and designation of barley chromosomes by Giemsa banding technique: A reconsideration. *Z. Pflanzenzüchtg.* 86: 336–340.

Singh, R. J., and T. Tsuchiya. 1982a. An improved Giemsa N-banding technique for the identification of barley chromosomes. *J. Hered.* 73: 227–229.

Singh, R. J., and T. Tsuchiya. 1982b. Identification and designation of telocentric chromosomes in barley by means of Giemsa N-banding technique. *Theor. Appl. Genet.* 64: 13–24.

Tjio, J. H., and A. Levan. 1950. The use of oxyquinoline in chromosome analysis. *An. Estacion Exp. Aula Dei.* 2: 21–46.

Tuleen, N. A. 1971. Linkage data and chromosome mapping, in *Barley Genetics II. Proc. 2nd. Int. Barley Genet. Symp.* R. A. Nilan, Ed., Washington State University Press, Pullman, pp. 208–212.

3 Smear Technique for Plant Chromosomes

3.1 INTRODUCTION

The removal of the cover glass is a prerequisite for Giemsa C- and N-banding techniques for plant chromosomes. To avoid this step, attempts are being made to prepare plant chromosome slides by a smear technique, as used for the mammalian chromosomes. Using a smear technique, chromosome slides have been prepared from the roots of several plant species (Singh 2017).

3.2 CHROMOSOME PREPARATION FROM THE ROOT

- Germination of seeds, collection, and pretreatment of roots should be conducted as described in Chapter 2.
- Cut only the meristematic region (1 mm) of the roots and treat roots with hypotonic 0.075 M KCl for 15–20 min at RT. However, it is not necessary to treat the roots with hypotonic solution.
- Fix roots in absolute ethanol/glacial acetic acid (3:1) or methanol/glacial acetic acid (3:1) for a minimum of 1–2 h or up to several months at –20°C.
- Rinse fixed roots in 0.1 M citric acid–sodium citrate buffer, pH 4.4–4.8.

3.2.1 ENZYME TREATMENT

Treatment of root tips by enzyme is quite variable:

- *Enzyme mixture 1*
  - 10% pectinase (Sigma P-5146)
  - 1.5% cellulase (Onozuka R-10)
  - Citrate buffer at 37°C for 30 min
- *Enzyme mixture 2*
  - 6% cellulase (Onozuka R-10)
  - 6% pectinase (Sigma)
  - Adjust pH 4.0 with HCl and treat roots for about 60 min at 35°C.
- *Enzyme mixture 3*
  - Pectinase, 20%–40% (v/v) (Sigma, from *Aspergillus niger*, P-5146, obtained as glycerol-containing stock solution)
  - Cellulase 2%–4% (w/v) (Calbiochem 21947 or Onozuka R-10)
  - 0.01 M citric acid–sodium citrate buffer

3.2.2 PREPARATION OF CHROMOSOME SPREAD

- *Flame dry method for rice*
  - Wash the roots in distilled water for 5–10 min at about 20°C to remove the enzyme.
  - Place root meristem on a clean slide with a drop of fresh fixative (3 parts methanol:1 part acetic acid).
  - Break the root meristem into fine pieces with a needle.
  - Add a few drops of the fixative and flame dry the slide.
- *Air-dry method for potato*
  - Transfer one root meristem to a clean slide.
  - Remove the excess buffer.
  - Add a drop of 60% acetic acid.
  - Heat slide (without boiling) over an alcohol flame and leave for 2–5 min.
  - Suspend cells in a drop of acetic acid with the help of a fine needle, leave for 1 min, and heat slightly again with tilting of the slide.
  - Add Carnoy's solution to suspension when slide is cooled down (some seconds), add three more drops of Carnoy's solution on the top of the suspension, air dry the slides, and store overnight or longer and stain as needed.
- *Air-dry method for white mustard*
  - Spin the cells at about 4000 rpm in small conical centrifuge tube (10 mL volume) at each step to change the solution.
  - Remove the supernatant carefully with a Pasteur pipette and resuspend the pellet in approximately 5 mL liquid.
  - Transfer fixed roots to buffer, collect only meristem tissues, and rinse twice in buffer.
  - Soften the tissue in enzyme solution at 37°C for 1–2 h depending upon the plant material.
  - Wash twice in buffer and finally suspend the pellet in a drop of buffer to prevent protoplasts sticking together.
  - Add an excess of fixative (freshly prepared) and change it twice, suspend the pellet in a small amount of fixative, drop this suspension onto ice-cold tilted slides and air dry.
  - Age air-dried slides overnight or longer before further processing.

Note: In addition to *Sinapis alba*, the above procedure produced an excellent chromosome spread of *Vicia faba, Pisum sativum, Crepis capillaris, Calla palustris, and Spirodela polyrrhiza* (Geber and Schweizer 1988).

3.3 CHROMOSOME PREPARATION FROM CELL SUSPENSION AND CALLUS

Chromosome count is conducted in cell suspension and calluses derived from the plants to determine if calluses/suspensions contain parental chromosome constitutions. If they contain chromosomal aberrations, they will not generate plants with parental chromosomes. This can be conducted by the squash method. However, Murata (1983) developed an air-dry method to determine chromosomes from suspension and callus cultures. The procedure has been divided into three steps: (1) pretreatment to accumulate metaphase cells, (2) cell wall digestion and protoplast isolation, and (3) application of air drying technique.

3.3.1 CHROMOSOME COUNT IN SUSPENSION CULTURE

- Add 1 mL of 0.5% colchicine solution to 9 mL of cell suspension 2–3 days after subculture in a 100 mm × 15 mm Petri dish and place on a gyratory shaker (50 rpm) for 2 h.
- Transfer 2 mL of cell suspension to 2 mL of enzyme solution in a 100 mm × 15 mm Petri dish (enzyme solution: 2% cellulysin [Calbiochem], 1% macerase [Calbiochem] and 0.6 M sorbitol; pH 5.5–5.6). Substitution of 1% pectolyase Y-23 (Kikkoman) for maceration produced comparatively faster protoplast isolation.
- Seal the Petri dish with parafilm and place on the gyratory shaker (50 rpm) for 3–4 h at RT (25°C).
- Filter the cells and enzyme mixture through 60 μm nylon mesh into a 15 mL centrifuge tube and centrifuge (65 ×g) for 3 min.
- Rinse twice with 0.6 M sorbitol and suspend in 5 mL hypotonic solution (0.075 M KCl), and allow to stand for 7 min at RT (25°C).
- Remove the supernatant following centrifugation (65 ×g) for 5 min, gradually add fresh fixative 3:1 (95% ethanol: glacial acetic acid) up to 5 mL, and allow to stand for 1 h.
- Resuspend in fresh fixative following centrifugation; repeat twice and make the final volume of the fixed cells 0.5–1.0 mL.
- Put five to six drops of the fixed cells by using a Pasteur pipette onto a wet cold slide and air or flame dry.
- Stain 3–4 min with 4% Giemsa (Gurr's R66, Bio/medical Special) diluted with 1/15 M phosphate buffer (pH 6.8), rinse in phosphate buffer and distilled water, and air dry.
- Mount slide in DePeX mounting medium (Bio/Medical Special).

3.3.2 CHROMOSOME COUNT IN CALLUS

- Put 10–20 mg calluses 5–7 days after subculture into a 15 mL centrifuge tube with 5 mL of Murashige and Skoog (1962) (MS) liquid medium containing 0.5% colchicine and allow to stand for 5 h.
- Discard the liquid medium following centrifugation (100 ×g) for 5 min and suspend in the fresh Carnoy's fixative (3:1) for 1 h.
- Rinse twice with dH$_2$O and add 5 mL of enzyme solution.
- Place the centrifuge tube, sealed with cap, horizontally on the gyratory shaker (100 rpm) for 2 h.
- Filter the cell suspension through 60 μm nylon mesh into another centrifuge tube.

- Rinse twice with 0.6 M sorbitol and suspend in a 5 mL hypotonic solution (0.075 M KCl) and allow to stand for 7 min at RT (25°C).
- Remove the supernatant following centrifugation (65 ×g) for 5 min, gradually add fresh fixative 3:1 (95% ethanol: glacial acetic acid) up to 5 mL, and allow to stand for 1 h.
- Resuspend in fresh fixative following centrifugation, repeat twice, and make the final volume of the fixed cells 0.5–1.0 mL.
- Using a Pasteur pipette, put five to six drops of the fixed cells onto a wet cold slide, and air or flame dry.
- Stain 3–4 min with 4% Giemsa (Gurr's R66, Bio/medical Special) diluted with 1/15 M phosphate buffer (pH 6.8), rinse in phosphate buffer and distilled water, and air dry.
- Mount slide in DePeX mounting medium (Bio/Medical Special).

For chromosome count in barley calluses, Singh (1986) used the following protocol:

- Pretreat actively growing morphogenic calluses in ice-cold water for 18 h.
- Fix calluses in 3:1 (95% ethanol:propionic acid).
- Wash calluses in 70% ethanol twice and store in a −20°C refrigerator.
- Stain a small portion of callus in 1% acetocarmine for 1 week.
- Take a very small part of stained callus, place onto the slide, and apply a drop of 45% acetic acid.
- Heat on a low flame of an ethanol burner, suspend callus tissue in 45% acetic acid, remove large debris, heat, add a drop of 45% acetic acid, and apply 18 mm × 18 mm × 1 mm cover glass.
- Reheat but not to boil, tap gently, apply slight pressure, and observe cells through a microscope.
- Cell with chromosome numbers were $2n = 7$ (Figure 3.1a), $2n = 14$ (Figure 3.1b), $2n = 28$, and $2n = 56$ (Figure 3.1c).

3.4   CHROMOSOME COUNT FROM LEAF EXPLANTS

3.4.1   POTATO

Pijnacker et al. (1986) studied the chromosomes of leaf explants of potato by following a smear technique:

- Pretreat in a saturated α-bromonaphthalene for 3 h at RT before fixation.
- Fix leaf explants directly in 3:1 (ethanol 100%:glacial acetic acid) fixative at 4°C for about 24 h.
- Rinse leaf pieces in distilled water.
- Incubate in 15% (v/v) pectinase (Sigma 5146) + 1.5% (w/v) cellulase R 10 (Yakult) in citrate buffer pH 4.8 for 45 min at 37°C.
- Rinse and then keep in distilled water for a minimum of 2 h.
- Transfer one leaf piece to a clean slide and add a drop of 60% acetic acid.
- Make leaf pieces into a fine suspension.

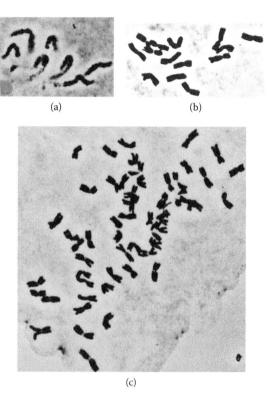

(a)   (b)

(c)

FIGURE 3.1 Chromosome count in morphogenic calluses of barley. (a) $2n = 7$, (b) $2n = 14$, and (c) $2n = 58$.

- Surround this suspension with cold fixative (3:1), then add about three drops of fixative on the top of the suspension, and air dry the slide.
- Stain slides in Hoechst 33258 (2 mg Hoechst in 1 mL ethanol and 0.3 mL of this solution in 100 mL 0.5× SSC) for 15 min at RT.
- Wash in 0.5× SSC, apply cover glass, and expose to 285–385 nm UV light at a distance of 10 cm for 15 min.
- Remove cover glass in 0.5× SSC and incubate in 2× SSC at 60°C for 30 min.
- Rinse slides in distilled water and stain in 2% Giemsa in Sörensen buffer, pH 6.9, for 30 min.
- Rinse slides in buffer and distilled water, air dry, dip in xylene, and mount in DePeX.
- Observe slides through a microscope (Figures 3.2 through 3.6).

3.4.2 BEETROOT

Devaux (1996) developed the following protocol for counting chromosomes of beetroot:

- Excise young leaves (a few mm) about 4 p.m.–5 p.m.
- Pretreat shoots in 1% bromonaphthalene for 2 h at RT.

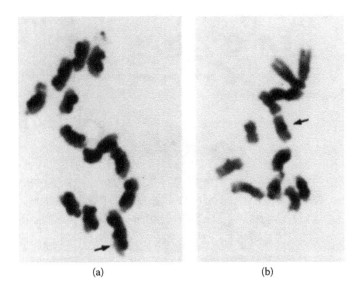

(a)   (b)

FIGURE 3.2 Chromosome count in monohaploid ($2n = \times = 12$) potato leaf explants. (a) Differential stained monochromosomes after two rounds of BrdC incorporation. (b) $2n = 12$ (right, after three rounds). (From Pijnacker, L.P., Walch, K., and Ferwerda, M.A., *Theor. Appl. Genet.*, 72, 833–839, 1986.)

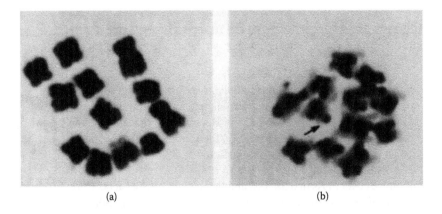

(a)   (b)

FIGURE 3.3 Chromosome count in monohaploid ($2n = \times = 12$) potato leaf explants. (a) Normally stained diplochromosomes. (b) Differentially stained diplochromosomes with two darkly stained inner chromatid after two rounds of BrdC incorporation. (From Pijnacker, L.P., Walch, K., and Ferwerda, M.A., *Theor. Appl. Genet.*, 72, 833–839, 1986.)

- Fix shoots in 90% acetic acid for 30 min at RT.
- Rinse twice in 95% ethanol.
- Transfer in 70% ethanol and store at 4°C up to several months.
- Wash twice in tap water.
- Hydrolyze in 5N HCl for 10 min at RT, then 1N HCl for 5 min at RT.

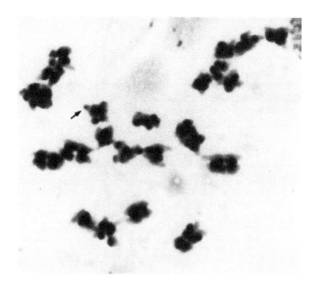

FIGURE 3.4 A tetraploid cell with $2n = 24$. (From Pijnacker, L.P., Walch, K., and Ferwerda, M.A., *Theor. Appl. Genet.*, 72, 833–839, 1986.)

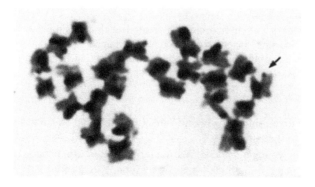

FIGURE 3.5 Differential stained diplochromosomes ($2n = 24$) with one darkly stained inner chromatid after three rounds of BrdC incorporation. (From Pijnacker, L.P., Walch, K., and Ferwerda, M.A., *Theor. Appl. Genet.*, 72, 833–839, 1986.)

- Stain shoots in Schiff's reagent for 1 h at RT.
- Squash 1/6 of a leaf of 2 mm length between slide and cover glass in one drop of iron acetocarmine.

3.4.3 ROSE

Ma et al. (1996) developed a cytological procedure to study mitotic metaphase chromosomes from shoots of roses.

- Collect actively growing terminal shoot (2–4 mm) in the morning during the spring time burst of growth.

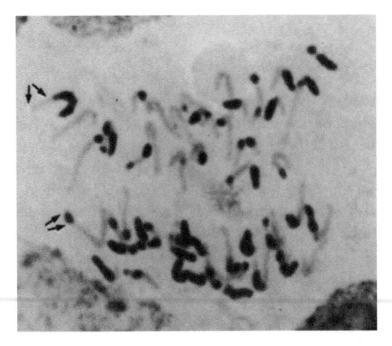

FIGURE 3.6 Behavior of chromosomes in potato leaf tissue cultured *in vitro* as studied by BrdC-Giemsa labeling; anaphase of differentially stained diplochromosomes ($2n = 24$). Arrows point to some of the sister chromatid exchanges. (From Pijnacker, L.P., Walch, K., and Ferwerda, M.A., *Theor. Appl. Genet.*, 72, 833–839, 1986.)

- Place five to ten shoot tips in one 5 mL tube and keep on ice water (0°C) for transport to the laboratory.
- Remove the young outside leaves, cut terminal 2–4 mm portion of the shoot apex and place immediately in a pretreatment solution (0.1% colchicine + 0.001 M 8-hydroxyquinoline) for 4 h in the dark at ≈25°C.
- Fix shoot tips in a *freshly prepared mixture* of 2 acetone:1 acetic acid (v/v) + 2% (w/v) polyvinylpyrrolidone (PVP) (MW 40,000).
- Fix tissues for 24 h at ambient pressure.
- Soak shoot tips in dH$_2$O for 1–24 h to elute the fixative and soften the tissues.
- Hydrolyze shoot tips in 1N HCl for 20 min at ≈25°C. Remove 1N HCl, wash twice in dH$_2$O, and soak for 10 min in dH$_2$O.
- Incubate shoot tips in 0.075 M KCl for 30–60 min. (Alternative: 0.01 M sodium citrate, pH 4.6, for 15–30 min).
- Digest the shoot tips in 5% cellulase R10 (Yakult Honsha Pharmaceutical, Tokyo) + 1% pectolyase Y23 (Seishin, Tokyo) + 0.01M sodium citrate at pH 4.6. Use 2–4 μL of enzyme mix per shoot tip. Digest shoot tip in enzyme for 3–4 h at 37°C.
- Prepare slides either by one-slide maceration or by suspension and spreading of protoplasts.

- *One -slide maceration*
  - Place one shoot tip on a microscope slide and blot excess water.
  - Add two to three drops of 3:1 or 1:1 (1 ethanol:1 acetic acid) fixative.
  - Disperse tissues gently with forceps tips and examine briefly at ×160, until most of the cells are settled onto the surface of the slide.
  - Wash cytoplasmic debris with additional drops of fixative, remove excess fixative by filter paper, air dry, stain, and cover with a cover slip.
- *Suspension and spreading of protoplast*
  - Add about 480 μL dH_2O to microcentrifuge tube containing the shoot tips and enzyme mix, vortex tube vigorously for 30–60 s to break up the shoot tips, and discard undissociated pieces.
  - Centrifuge at $\approx 700g_n$ fixed suspension protoplast for 5 min at 2°C–4°C and discard the supernatant.
  - Resuspend the pellet in ≈ 500 μL dH_2O and transfer supernatant cell suspension to a new microcentrifuge tube.
  - Centrifuge the suspension at $\approx 700g_n$ for 5 min at 2°C–4°C, discard the supernatant, and resuspend pellet in ≈ 500 μL of 3:1 (ethanol:acetic acid) fixative.
  - Centrifuge the suspension at $\approx 700g_n$ for 5 min at 2°C–4°C, discard the supernatant and resuspend the pellet in 15–20 μL freshly made 3:1 (ethanol:acetic acid) fixative per shoot tip.
  - Apply 5–10 μL of the suspension to a scrupulously clean microscope slide and allow it to air dry ($\approx 30\%$–50% relative humidity).
  - Allow suspension to spread out and air dry without further disturbance.
  - Apply a drop of stain and a cover glass to the dried slide.
  - Heat to 87°C for ~20 s and squash on a hot plate at 87°C.
  - Observe chromosome slide by microscope.
- For FISH, allow slides to air dry for 2–3 days at low humidity to improve cell retention.

3.4.4 BANANA

This protocol for meiotic chromosomes for banana and plantain was described by Adeleke et al. 2002. Cytological examination of these plants is difficult because chromosomes are small karyologically.

- *Selection of anthers*: Collect male buds on sunny days between 11:30 a.m. and noon
- Remove bracts carefully to expose the cluster of male flowers
- Dissect out anthers from the floral buds and examine the stage of meiosis by acetocarmine stain
- *Fixation of cells and enzyme digestion*: Fix buds in 3 ethanol:1 acetic acid + 1% ferric chloride for 18–24 h at 4°C
- Place five to six anthers side-by-side on a glass slide and cut off both ends with a sharp razor blade

- To prevent the anthers from drying, cover anthers in a few drops of LB01 buffer (15 mM Tris, 2 mM NaEDTA, 80 mM KCl, 0.5 mM spermine, 15 µM mercaptoethanol, 0.1% Triton X-100, pH 7.5; Doležel et al. 1998)
- Squeeze out both ends of the anthers to release microsporocytes with a dissecting needle
- Transfer the clumps of microsporocyte suspension to a microcentrifuge tube and centrifuge at 1000 ×g for 5 min
- Discard the supernatant and resuspend the pellet in cold citrate buffer (0.1 M sodium citrate and 0.1 M citric acid, pH 4.5) and repeat the process
- Resuspend the pellet in 0.2 mL of an enzyme mixture (5% cellulose, Sigma Chemicals; 1% pectinase and 1% pectolyase Y23, Karlan Research, Santa Rosa, California)
- Prepare enzyme in 0.01 M sodium citrate buffer consisting of 0.01 M sodium citrate and 0.01 M citric acid, pH 4.5
- Incubate at 37°C for 30–60 min
- Pellet cells from the enzyme solution, wash by centrifugation in citrate buffer, and finally resuspend in ice-cold 70% ethanol
- *Slide preparation*: Place a drop of suspension on a clean slide with a Pasteur pipette and air dry
- Apply a drop of freshly prepared 3:1 (ethanol:acetic acid) fixative over the cells
- Examine the quality of the slide by phase contrast microscope
- Air dry slide and stain with freshly mixed Giemsa stain (Sigma USA)-3mL stock solution of Giemsa + 60 mL Sörenson phosphate buffer (30 mL KH_2PO_4 + 30 mL $Na_2PO_4.H_2O$, pH 6.9)—Leishman's stain (BDH, UK) dilute in a phosphate buffer at pH 6.8 for 30–45 min
- For silver nitrate staining, immerse slides in 2×SSC (0.3 M NaCl, 0.3 M Na citrate, pH 7.0) for 10 min at RT
- Rinse briefly in distilled water and air dry
- Add two drops of a 100% silver nitrate aqueous solution over the cells
- Apply a plastic coverslip to spread the stain over the slide
- Incubate slides for 2–4 min at 60°C in a humid chamber
- Rinse slides few times in distilled water and air dry
- Dip slides in xylene for 30 min and make permanent by mounting in DPX
- Examine slides in a Leitz Diaplan light microscope
- Photograph well-spread chromosome cell using 100× 1.25 oil immersion objective

3.4.5 COFFEE

Clarindo and Carvalho (2006) developed a chromosome preparation protocol for coffee (*Coffea canephora*; $2n = 2x = 22$) suspension cells. This species has a relatively small size of chromosomes and the mitotic index from root meristem is low.

- Add microtubule inhibiting agent AMP (amiprophos-methyl) to the culture media to a final concentration of 3 µM, for a period of 2, 3, 4, 5, 6, 12, and 24 h
- Fix cell aggregates in methanol:acetic acid (3:1)

- Change fixative three times at 10-min intervals
- Store cell aggregates at −20°C from one to several days
- Wash cell aggregates and digest with enzymatic solution flaxzyme (Novo Ferment™) and distilled water in a ratio of 1:30 (enzyme:distilled water) for 30 min at 34°C
- Wash the aggregates for 10 min in distilled water.
- Fix again and store at −20°C
- Prepare slides by cell dissociation of the enzymatic macerated aggregates and subsequently air dry on a hot plate at 50°C
- Stain slides with 5% Giemsa solution (Merck) in phosphate buffer (pH 6.8) for 5 min
- Wash slides twice in distilled water and air dry

3.4.6   Ag-NOR Banding

This technique was modified from Goodpasture and Bloom (1975).
Prepare two solutions:

1. Pretreatment (Ag) solution of 50% aqueous silver nitrate.
2. *An ammoniacal silver (AS) solution*: Prepare by dissolving 4 g silver nitrate in a solution of 5 mL distilled water and 5 mL concentrated ammonium hydroxide (pH 12–13).
   a. Neutralize the developing solution of 3% formalin first with sodium acetate crystals and then adjust pH 5–6 with formic acid.
   b. Pipette Ag solution onto a slide, cover with a cover glass and place slides about 25 cm below a photo flood (2800 K bulb) for 10 min.
   c. Rinse off cover glass in distilled water and develop slides by adding four drops of AS solution followed by four drops of developing solution.
   d. Immediately cover the slide with a cover glass and monitor the progress of the stain under the microscope.
   e. When the chromosomes reach a golden yellow color, rinse slide in distilled water, dehydrate in an ethanol series, soak in xylene, and mount in permount.
   f. Apply 50% $AgNO_3$ solution on the slide.
   g. Incubate at 34°C for 18 h.
   h. Remove cover glass and wash slides in distilled water for 2 min.

3.4.7   Hsc-FA Banding

Destain Giemsa-stained slides with methanol:acetic acid (3:1) solution.

- Incubate aged slides for 7 days in phosphate buffer (pH 6.5) at 85°C for 20 min.
- Stain with 0. 01% (w/v) AO solution for 15 min.
- Wash slides in distilled water for 2 min and phosphate buffer for 3 min.
- Add three drops of buffer to the slides and cover by cover glass.

3.4.8   IMAGE ANALYSIS

Capture chromosome images with CoolSNAP-Pro *cf* (Roper Scientific) camera on an Olympus™ BX-60 fluorescence microscope with a 100× objective lens and WB filter.

- Digitize frame using Image PRO®-Plus analysis system (Media Cybernetics).
- Perform image analysis on a Power Macintosh G4 computer using the freely available (http://reg.ssci.liv.ac.uk) Image SXM software.
- Cell aggregates treated with 3 µM APM for 3 or 4 h produce well-defined primary and secondary constrictions and help in generating a karyogram (Figure 3.7) that shows the presence of secondary constriction on the distal short arm of chromosome 6 (Figure 3.7a through d). An active NOR by Ag-NOR banding can be shown in Figure 3.8.
- Hsc-FA banding shows a uniformly red–orange fluorescence for all chromosomes and a strong yellowish-green fluorescence band on the secondary constriction of chromosome 6 in metaphase (Figure 3.7a) and with more pronounced flanking NOR in the mitotic prophase (Figure 3.8b).

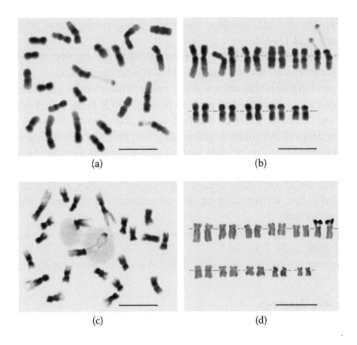

(a)   (b)

(c)   (d)

FIGURE 3.7 Chromosomes from cell suspension aggregates culture of *Coffea canephora* treated with 3 µM APM for 3 h. (a) Metaphase chromosome stained with 5% Giemsa, (b) karyogram ($2n = 22$) showing distinct morphology of 11 pairs of chromosomes, (c) metaphase chromosomes stained with Giemsa showing the nucleoli-associated NOR of chromosome 6, (d) the karyogram by Ag-NOR staining confirming that chromosome 6 is a NOR. (From Clarindo, W.R., and Carvalho, C.R., *Cytologia,* 71, 243–249, 2006.)

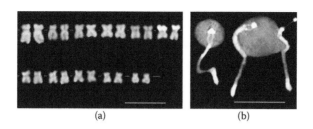

(a)   (b)

FIGURE 3.8 (a) Metaphase chromosomes of *C. canephora* stained by Hsc-FA showing one band emitting yellowish-green fluorescence at the subterminal position of chromosome 6 and (b) chromosome 6 associated with the nucleoli. (From Clarindo, W.R., and Carvalho, C.R., *Cytologia,* 71, 243–249, 2006.)

3.5 CHROMOSOME PREPARATION FROM FLOWERS

Murata and Motoyoshi (1995) developed a cytological procedure for *Arabidopsis thaliana* floral tissues of young buds.

- Dissect out sepals from young buds (1.5–2 mm in length) under a dissecting microscope.
- Transfer buds without sepals into a 1.5 mL micro tube with 1 mL dH_2O, keep at 0°C in iced water for 24 h to allow accumulation of metaphase cells.
- Fix buds in 3:1 (99% methanol: glacial acetic acid), and store at −20°C.
- For slide preparation, rinse buds well with dH_2O.
- Digest buds with an enzyme solution containing 2% (w/v) cellulase Onozuka R10 (Kinki Yakult) and 20% (v/v) pectinase (Sigma) and incubate for 1 h 30 min at 30°C.
- Suspend tissues with micropipette, rinse suspended cells with dH_2O, and fix again.
- After three changes of fresh fixative, drop fixed cells onto wet, cold slides and flame dry.
- Stain slides with 4% (v/v) Giemsa solution diluted with 1/15 M phosphate buffer, pH 6.8 for 20–30 min, and examine with a microscope (Figure 3.9).

3.5.1 MEIOTIC CHROMOSOMES

- Collect barley spikelets still covered with the flag leaf.
- Bring spikelets into the laboratory, check meiotic stage by dissecting out one anther.
- If desired meiotic stage is identified, fix remaining anthers in 3:1 (95% ethanol:glacial acetic acid).
- Fixation in 3:1 (95% ethanol: propionic acid) with 1 g ferric chloride per 100 mL fixative produces meiotic cells with stained chromosomes and clear cytoplasm (Figure 3.10).
- Wash anthers in 70% ethanol twice and store at −20°C.
- Stain roots in 1% propionic acid carmine for 1 week in a refrigerator.

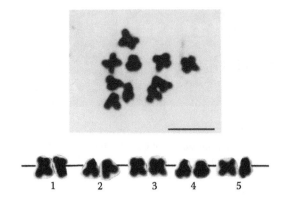

FIGURE 3.9 Floral chromosomes of *Arabidopsis thaliana* prepared from cold-treated young flower buds and its karyotype. (From Murata, M., and Motoyoshi, F., *Chromosoma,* 104, 39–43, 1995.)

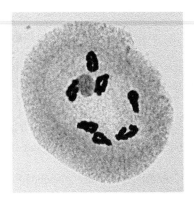

FIGURE 3.10 A meiotic pollen mother cell of barley at the diakinesis stage showing seven bivalents; two pairs are associated with nucleolus.

- Squeeze out pollen mother cells from the anther in 45% acetic acid, apply 18 mm × 18 mm × 1 mm cover glass, heat on a very low flame, and remove excess 45% acetic acid by a filter paper.
- Observe cells using a microscope.

REFERENCES

Adeleke, M. T. V., M. Pillay, and B. E. Okoli. 2002. An improved method for examining meiotic chromosomes in *Musa L. Hort. Sci.* 37: 959–961.

Clarindo, W. R., and C. R. Carvalho. 2006. A high quality chromosome preparation from cell suspension aggregates culture of *Coffea canephora. Cytologia* 71: 243–249.

Devaux, P. 1996. Number and morphology of chromosomes –leaf meristem. *In Techniques of Plant Cytogenetics.* J. Jahier (Ed.), Science Publishers, Inc. Lebanon, NH.

Doležel, J., J. Greilhuber, S. Lucretti, A. Meister, M. A. Lysak, L. Nardi, and R. Obermayer. 1998. Plant genome size estimation by flow cytometry: Inter-laboratory comparison. *Ann. Bot.* 82(A): 17–26.

Geber, G., and D. Schweizer. 1988. Cytochemical heterochromatin differentiation in Sinapsis alba (Cruciferae) using a simple air-drying technique for producing chromosome spreads. *Pl. Syst. Evol.* 158: 97–106.

Goodpasture, C., and S. E. Bloom. 1975. Visualization of nucleolar organizer regions in mammalian chromosomes using silver staining. *Chromosoma (Berl.)* 53: 37–50.

Ma, Y., M. N. Islam-Faridi, C. F. Crane, D. M. Stelly, H. J. Price, and D. H. Byrne. 1996. A new procedure to prepare slides of metaphase chromosomes of roses. *Hort. Sci.* 31: 855–857.

Murata, M. 1983. Staining air dried protoplasts for study of plant chromosomes. *Stain Tech.* 58: 101–106.

Murata, M., and F. Motoyoshi. 1995. Floral chromosomes of *Arabidopsis thaliana* for detecting low-copy DNA sequence by fluorescence in situ hybridization. *Chromosoma* 104: 39–43.

Murashige, T., and F. Skoog. 1962. A revised medium for rapid growth and bioassays with tobacco tissue cultures. *Physiol. Plant.* 15: 473–497.

Pijnacker, L. P., K. Walch, and M. A. Ferwerda. 1986. Behaviour of chromosomes in potato leaf tissue cultured in vitro as studied by BrdC-Giemsa labelling. *Theor. Appl. Genet.* 72: 833–839.

Singh, R.J. 1986. Chromosomal variation in immature embryo derived calluses of barley (*Hordeum vulgare L.*). *Theor. Appl. Genet.* 72: 710–716.

Singh, R. J. 2017. Plant Cytogenetics, Third Edition. CRC Press (Taylor & Francis Group). Boca Raton, FL.

4 Fluorescence *In Situ* Hybridization

4.1 INTRODUCTION

Fluorescence *in situ* hybridization (FISH) is a very powerful technique for the detection of specific nucleic acid sequences and to localize highly repetitive DNA sequences in the specific regions of the chromosomes. Since the publication of Gall and Pardue (1969), numerous modifications and refinements of techniques have facilitated the diagnosis and identification of chromosomal aberrations particularly for human and animal chromosomes.

The application of an *in situ* hybridization technique in plants has lagged behind compared to its use in mammalian cytogenetics. The main handicap of utilizing this technique for plant chromosomes is to obtain a high frequency mitotic metaphase cells without cell wall and cytoplasmic debris. These obstacles hinder hybridization of low-copy-number sequences to the chromosomes. The *in situ* hybridization technique has been used to identify chromosomes in several plant species, particularly in wheat and its allied genera. Singh et al. (2001) developed a procedure to detect repetitive DNA sequences on somatic metaphase chromosomes for soybean ($2n = 40$) root tips using fluorescence *in situ* hybridization after consulting many published results for the mammalian as well as plant chromosomes.

4.2 EQUIPMENT

- Premium microscope slides, frosted, 25 mm × 75 mm × 1 mm (Fisher Scientific, Cat. no. 12-544-2).
- Premium cover glass, 18 mm × 18 mm × 1 mm (Fisher Scientific, Cat. no. 12-548-A).
- Microscope with 100×, 63× (oil), 40×, 20× objectives. Fluorescence light source.
- Computer attached to the microscope.
- Incubator (water bath and oven).
- Moist chamber (a plastic rectangle box with tight-fitting lid; this maintains the moisture atmosphere). The bottom of the chamber is covered with the paper towel moist with 2×SSC. Two glass rods are kept on the moist paper towel, and slides are placed on the glass rod.

4.3 DNA ISOLATION

This protocol for DNA extraction was developed by modifying protocol III by Jeff and Jane Doyle published first in *Phytochemical Bulletin* in 1987 and again in *Focus*

(12:13–15, 1990). The DNA extraction procedure, described below, was kindly provided by Jeff Doyle (March 4, 2016).

4.3.1 Buffer/Tissue

- CTAB isolation buffer (2% hexadecyltrimethylammonium bromide; CTAB; Sigma H-5882)
- 1.4 M NaCl
- 0.2% 2-mercaptoethanol
- EDTA 100 mM Tris-HCl, pH 8.0)
- Glass centrifuge tube

4.3.2 Procedure

- Preheat 5–7.5 mL of CTAB isolation buffer (2% hexadecyltrimethylammonium bromide [CTAB: Sigma H-5882]), 1.4 M NaCl, 0.2% 2-mercaptoethanol, 20 mM EDTA, 100 mM Tris-HCl, pH 8.0, in a 30 mL glass centrifuge tube to 60°C in a water bath.
- Grind 0.5–1.0 g fresh leaf tissue in 60°C CTAB isolation buffer in a preheated mortar.
- Incubate sample at 60°C for 30 min (15–60 min) with optional occasional gentle swirling.
- Extract once with chloroform–isoamyl alcohol (24:1), mixing gently but thoroughly.

This produces two phases, an upper aqueous phase that contains the DNA and a lower chloroform phase that contains some degraded proteins, lipids, and many secondary compounds. The interface between these two phases contains most of the "junk" cell debris, many degraded proteins, etc.

- Spin in clinical centrifuge (swinging bucket rotor) at room temperature to concentrate phases. We use setting 7 on our IEC clinical (around 6000 × g) for 10 min.

This is mainly to get rid of the junk that is suspended in the aqueous phase. Generally the aqueous phase will be clear, though often colored, following centrifugation, but this is not always the case.

- Remove aqueous phase with wide bore pipette, transfer to clean glass centrifuge tube, add 2/3 volumes cold isopropanol, and mix gently to precipitate nucleic acids.
- A wide bore pipette is used because DNA in solution is a long, skinny molecule that is easily broken (sheared) when it passes through a narrow opening. Gentleness also improves the quality (length) of DNA. In some cases, this stage yields large strands of nucleic acids that can be spooled out

with a glass hook for subsequent preparation. Mostly, this does not happen, however, and the sample is flocculent, merely cloudy-looking, or, in some instances, clear. If no evidence of precipitation is observed at this stage, the sample may be left at room temperature for several hours or overnight. This is one convenient stopping place, in fact, when many samples are to be prepped. In nearly all cases, there is evidence of precipitation after the sample has been allowed to settle out in this manner.

- If possible, spool out nucleic acids with a glass hook and transfer to 10–20 mL of wash buffer (76% ethanol, 10 mM ammonium acetate).
  - *Preferred alternative*: Spin in clinical centrifuge (e.g., setting 3 on IEC) for 1–2 min and gently pour off as much of the supernatant as possible without losing the precipitate, which will be a diffuse and very loose pellet. Add wash buffer directly to pellet and swirl gently to resuspend the nucleic acids.
  - *Last resort*: Longer spins at higher speeds may be unavoidable if no precipitate is seen at all. This will result, generally, in a hard pellet (or, with small amounts, a film on the bottom of the tube) that does not wash well and may contain more impurities. Such pellets are difficult to wash, and in some cases, we tear them with a glass rod to promote washing at which point they often appear flaky.

Nucleic acids generally become much whiter when washed, though some color may still remain.

- Spin down (or spool out) nucleic acids (setting 7 IEC, 10 min) after at least 20 min of washing. The wash step is another convenient stopping point, as samples can be left at room temperature in a wash buffer for at least 2 days without noticeable problems.
- Pour off supernatant carefully (some pellets are still loose even after this longer spin) and allow to air dry briefly at room temperature.
- Resuspend nucleic acid pellet in 1 mL TE (10 mM Tris-HCl, 1 mM EDTA, pH 7.4).

Although we commonly continue through additional purification steps, DNA obtained at this point is generally suitable for restriction digestion and amplification, so we'll stop here.

If DNA is to be used at this stage, pellets should be more thoroughly dried than indicated above.

Gel electrophoresis of nucleic acids at this step often reveals the presence of visible bands of ribosomal RNAs as well as high molecular weight DNA.

- Add RNase A to a final concentration of 10 μg/mL and incubate 30 min at 37°C.
- Dilute sample with two volumes of distilled water or TE, add ammonium acetate (7.5 M stock, pH 7.7) to a final concentration of 2.5 M, mix, add 2.5 volumes of cold EtOH, and gently mix to precipitate DNA.

DNA at this stage usually appears cleaner than in the previous precipitation. Dilution with water or TE is helpful, as we have found that precipitation from 1 mL total volume often produces a gelatinous precipitate that is difficult to spin down and dry adequately.

- Spin down DNA at high speed (10,000 × g for 10 min in refrigerated centrifuge, or setting 7 in clinical for 10 min).
- Air dry sample and resuspend in appropriate amount of TE.

4.4   NICK TRANSLATION

4.4.1   NICK TRANSLATION SYSTEM

This protocol is from the Bethesda Research Laboratory.

- Add the following to a 1.5 mL microcentrifuge tube on ice: 5 μL solution A4 (no dTTP; containing dATP, dCTP, dGTP, 0.2 mM each), × μL with sterile redistilled water and mix briefly.
- Add 5 μL solution C (DNA polymerase I and DNase I), mix gently but thoroughly, and centrifuge briefly (microcentrifuge 5000 rpm for 5s).
- Incubate at 15°C for 1 h.
- Add 5 μL of solution D (stop buffer).

4.4.2   NICK TRANSLATION KIT

This protocol is from Boehringer Mannheim.

- Add the following to a 1.5 mL microcentrifuge tube on ice: 5 μL of 0.4 mM dATP (vial 2), 5 μL of 0.4 mM dCTP (vial 3), 5 μL of 0.4 mM dGTP (vial 4), 2 μL of 10 mM biotin-16-dUTP, × μL of probe DNA (0.1–2.0 μg), 5 μL of 10 × nick translation buffer (vial 6). Make up to 45 μL with sterile redistilled water.
- Add 5 μL of enzyme mixture (DNA polymerase I and DNase I, vial 7).
- Incubate at 15°C for 90 min.
- Stop the reaction by heating to 65°C for 10 min.

4.4.3   LABELING REACTION WITH DIG-dUTP OR BIOTIN-dUTP

- Place a 1.5 mL microcentrifuge tube on ice and add to the tube 16 μL sterile redistilled H$_2$O containing 1 μg template DNA (either linear or supercoiled) and 4 μL of either DIG nick translation mix for *in situ* probes or Biotin nick translation mix for *in situ* probes.

Note: Each nick translation mix contains 5× concentrated reaction buffer; 50% glycerol; DNA Polymerase I; DNase I; 0.25 mM each of dATP, dCTP, and dGTP; 0.17 mM dTTP; and 0.08 mM X-dUTP (X = DIG or biotin).

1. Mix ingredients and centrifuge tube briefly.
2. Incubate at 15°C for 90 min.

3. Chill the reaction tube to 0°C.
4. Take a 3 μL aliquot from the tube and analyze it as follows:
   a. Mix the aliquot with enough gel loading buffer to make a sample that will fit in one well of an agarose minigel.
   b. Denature the sample (DNA aliquot + gel loading buffer) for 3 min at 95°C.
   c. Place the denatured sample on ice for 3 min.
   d. Run the sample on an agarose minigel with a DNA molecular weight marker.
5. Depending on the average size of the probe, do one of the following:
   a. If the probe is between 200 and 500 nucleotides long, go to step 4c.
   b. If the probe is longer than 500 nucleotides, incubate the reaction tube further at 15° C until the fragments are of the proper size.

Note: If the fragment is too long, the labeled probe can also be sonicated to the proper size.

- Stop the reaction as follows:
  - Add 1 μL 0.5 M EDTA (pH 8.0) to the tube.
  - Heat the tube to 65°C for 10 min.

4.4.4 LABELING REACTION WITH FLUORESCEIN DUTP, AMCA-DUTP, OR TETRAMETHYL-RHODAMINE-DUTP

1. Prepare 50 μL of a 5× fluorophore labeling mixture (enough for about 12 labeling reactions) by mixing the following in a 1.5 mL microcentrifuge tube on ice:
   - 5 μL 2.5 mM dATP
   - 5 μL 2.5 mM dCTP
   - 5 μL 2.5 mM dGTP
   - 3.4 μL 2.5 mM dTTP
   - 4 μL of either 1 mM fluorescein-dUTP or 1 mM AMCA-dUTP or 1 mM tetramethylrhodamine-dUTP
   - 27.6 μL sterile redistilled H_2O
2. Place a 1.5 mL microcentrifuge tube on ice and add to the tube:
   - 12 μL sterile redistilled H_2O containing 1 mg template DNA (either linear or supercoiled)
   - 4 μL 5×concentrated fluorophore labeling mixture (from step 1 in the subsection 4.4.4)
   - 4 μL of nick translation mix for *in situ* probes

Note: Nick translation mix contains 5× concentrated reaction buffer; 50% glycerol; DNA polymerase I; and DNase I.

- Mix, incubate, and stop the reaction as in steps 2–7 for DIG and biotin labeling above.

4.4.5 PURIFICATION OF LABELED PROBE

1. Precipitate the labeled probe (from either procedure above) by performing the following steps:

   As alternative to the ethanol precipitation procedure below, you may purify labeled probes which are 100 bp or longer with the high pure PCR product purification kit.

   - To the labeled DNA, add 2.5 µL 4 M LiCl and 75 µL prechilled (–20°C) 100% ethanol, and mix well.
   - Let the precipitate form for at least 30 min at –70°C or 2 h at –20°C.
   - Centrifuge the tube (at 13,000 × g) for 15 min at 4°C.
   - Discard the supernatant.
   - Wash the pellet with 50 µL ice-cold 70% (v/v) ethanol.
   - Centrifuge the tube (at 13,000 × g) for 5 min at 4°C.
   - Discard the supernatant.
   - Dry the pellet under vacuum.

CAUTION: Drying the pellet is important because small traces of residual ethanol will cause precipitation if the hybridization mixture contains dextran sulfate. Trace ethanol can also lead to serious background problems.

2. Do one of the following:

   - If you are not going to use the probe immediately, dissolve the pellet in a minimal amount of TE (10 mM Tris-HCl, 1 mM EDTA, pH 8.0) buffer and store the probe solution at –20°C.

CAUTION: Avoid repeated freezing and thawing of the probe.

If you are going to use the probe immediately, dissolve the pellet in a minimal amount of an appropriate buffer (such as TE or sodium phosphate), then dilute the probe solution to a convenient stock concentration (e.g., 10–40 ng/µL) in the hybridization buffer to be used for the *in situ* experiment.

4.4.6 NICK TRANSLATION LABELING OF DOUBLE-STRANDED DNA WITH DIG-, BIOTIN-, OR FLUOROCHROME-LABELED dUTP

Note: For *in situ* hybridization procedures, the length of the labeled fragments obtained from this procedure should be about 200–400 bases.

1. Place a 1.5 mL microcentrifuge tube on ice and add to the tube:
   - 27 µL sterile, redistilled H_2O
   - 5 µL 10× concentrated nick translation buffer (500 mM Tris-HCl [pH 7.8], 50 mM $MgCl_2$, and 0.5 mg/mL bovine serum albumin [nuclease-free])

- 5 μL 100 mM dithiothreitol
- 4 μL nucleotide mixture (0.5 mM each of dATP, dGTP, and dCTP; 0.1 mM dTTP)
- 2 μL 1 mM DIG-dUTP or 1 mM biotin-dUTP or 1 mM fluorochrome-labeled dUTP
- 1 μg template DNA
- 5 μL (5 ng) DNase I
- 1 μL (10 units) DNA polymerase I

4.5 PREPARATION OF BUFFERS

4.5.1 PREPARATION OF CA–SC BUFFER

- M citric–sodium citrate buffer pH 4.5–4.8 (CA–SC)
- Prepare from 10× stock solution consisting of 3 parts 0.1 M trisodium citrate and 2 parts 0.1 M citric acid, check pH. It should be in range of 4.5–4.8.

4.5.2 PREPARATION OF SSC BUFFER

- 20× SSC
- 70.12 g NaCl
- 52.8 g Na citrate
- 4L ddH$_2$O, mix, filter, sterilize, and adjust pH 7.4. (stock)

4.5.3 PREPARATION OF 4× SSC + 0.2% TWEEN 20

- 400 mL 20× SSC
- 1596 mL ddH$_2$O
- 4 mL Tween 20

4.5.4 PREPARATION OF 2× SSC

- 100 mL 20× SSC
- 900 mL ddH$_2$O

4.6 PREPARATION OF ENZYME SOLUTION

- 0.02 g cellulase (2%) (cellulase Onozuka R-10 from Yakult, Tokyo, Japan)
- 0.01g pectinase (1%) (pectinase, Sigma, Cat. # P-2401)
- Dissolve in 1.0 mL 0.01 M CA–SC buffer in an Eppendorf tube

4.7 PREPARATION OF SLIDES FOR SOYBEAN CHROMOSOMES

For cereals, particularly wheat and barley, roots are pretreated in ice-cold water, fixed in Carnoy's Solution 1, stained in 1% acetocarmine. Slides are prepared by the squash method and the quality of chromosome spread is examined under a phase contrast lens. However, the roots of some plants are digested in the enzyme solution and slides are prepared by an air-dry method, described briefly as follows (Singh 2017):

1. Wash root tips three times, 5 min each in 0.01 M CA–SC buffer in a spot plate or in 1.5 mL Eppendorf tube at RT.
2. Place a Petri cover on some crushed ice and dissect only 1–2 mm (cream colored at the very tip) in length from the meristematic region of the root tips.
3. Transfer the tips into an Eppendorf tube with 0.01 M CA–SC buffer on ice.
4. Spin down in a microfuge for a few seconds and withdraw the supernatant.
5. Add 1.0 mL of enzyme solution, seal with Parafilm, and incubate in a 37°C oven/water bath for 1 h 30 min.
6. Aspirate the tips in the enzyme solution by drawing them into a Pasteur pipette several times. The tips should break apart and the solution should appear cloudy.
7. Again, incubate in a 37°C oven/water bath for 10–15 min.
8. Spin down as stated above.
9. Add about a 1 mL 0.01 M CA–SC buffer and let it sit on ice for 35–45 min.
10. Spin (3000 rpm, 3 min), resuspend the pellet in 75 mM KCl (150 mM KCl has also produced good results), and let it sit for 10 min at RT.
11. Spin (5000 rpm, 3 min), suspend in fresh fixative (3 parts 100% ethanol: 1 part glacial acetic acid), and let it stand 17–20 min at RT.
12. Spin (5000 rpm, 3 min) and suspend pellet in approximately 20–25 drops of fixative from a Pasteur pipette.
13. Place one drop of this suspension per slide using a Pasteur pipette. *Note:* Use slides that are ice-cold from a −80°C freezer. Drop from about 3 cm above on the flat slide and blow very gently on the slide to aid spreading.
14. Air dry slides. Once dried, they can be stored in a slide box sealed in a bag at −80°C for several weeks.

4.7.1 PROCESSING SLIDES THROUGH FISH

1. Check slides for good metaphase cells by mounting in 0.2 μg/mL Sigma DAPI in 1× PBS, pH 7.0, or in 0.4 μg/mL Sigma propidium iodide in 1× PBS, pH 7.0. Mark the position of cells from the scale on the microscope stage.
2. Place the slides in a −80°C freezer for 5 min and remove the cover glass with a razor blade.
3. Destain the slides by immersing in 45% acetic acid for 1–2 s, rinse in water, and air dry.
4. Scan slides for prometaphase and metaphase spreads by a phase contrast lens.
5. Remove cover slip after dipping slides in liquid nitrogen for a few seconds.

6. Cover slips can be also removed by dry ice.
7. Prepare eight slides.

4.7.1.1 Pre-hybridization Method

1. Treat slides in a staining jar in 45% acetic acid for 10 min at RT and air dry for 1–2 h or overnight at RT.
2. Turn on water bath at 70°C and prepare humid chamber.
3. Take 24 μL (from 1 mg/mL stock) RNase stock solution + 776 μL 2× SSC (800 mL). Add 100 μL RNase to each slide and cover with colored plastic cover slip (24 mm × 30 mm). Incubate at 37°C for 1 h in a humid chamber, drain slides (**Caution!** Slide should not get dry).
4. Treat slides in 70% formamide for 2 min at 70°C and drain slides.
5. Dehydrate slides at −20°C in 70% (5 min), 80% (5 min), 95% (5 min), and 100% (30 s) ethanol. Air dry at RT for 1–2 h.

4.7.1.2 Hybridization

1. Probe (50 μL)
   200 ng rDNA; Internal transcribed spacer (ITS) region of nuclear ribosomal DNA probe (ITS1, 5.8S, ITS2; approximately 700 nucleotides) (5 μL) + 19 μL dH$_2$O sterilized H$_2$O.
   a. Denature on PCR at 95°C for 5 min or in boiling water.
   b. Keep on ice 5 min (immediately).
   c. Add 5 μL 10× dNTP + 20 μL 2.5× random primer solution.
   d. Mix by tapping on the bottom.
   e. Add 10 μL Klenow from the freezer.
   f. Mix by tapping on the bottom.
   g. Incubate at 37°C for 3 h (Bio Prime™ DNA Labeling System, Life Technologies, Cat. no.: 18094-011, Lot no. EHDOO1).
2. Preparation of hybridization solution
   a. Each slide needs 50 μL of solution.
      i. 200 μL 50% formamide
      ii. 80 μL 50% dextron sulfate
      iii. 53 μL TE buffer
      iv. 40 μL 20× SSC
      v. 17 μL salmon sperm (ss) DNA
      vi. 10 μL probe
   b. Mix hybridization solution in a 1.5 mL tube, give it a quick spin, and transfer into three PCR tubes (two containing 150 μL and the third containing 100 μL).
   c. Denature at 80°C for 10 min (during this break prewarm humidity chamber), keep on ice for 10 min, give it a quick spin, mix, and keep on ice.
   d. Apply 50 μL denatured hybridization solution on each slide, cover, and keep at 80°C for exactly 8 min. Move humid chamber with slides quickly to 37°C overnight for hybridization (at this time, prepare 4× SSC + Tween 20 [will be known as 4× SSC] and 2× SSC)

3. *Posthybridization (next day)*
   a. Turn on the water bath (40°C), place two Coplin jars (designated as jar) with 2× SSC, one with 35 mL 2× SSC, one tube containing 35 mL formamide, and one bottle with 400 mL 2× SSC.
   b. Remove cover slip and drain slides.
   c. Wash slides in 2× SSC at 40°C for 5 min twice (discard used 2× SSC and fill with fresh 2× SSC). (During the second wash, prepare 5% bovine serum albumin (BSA): dissolve 0.24 g albumin bovine in 4.8 mL 4× SSC.) Drain slides.
   d. Treat slides in 50% formámide for 10 min at 40°C and drain.
   e. Wash in 2× SSC for 5 min twice at 40°C, wash third time in 2× SSC at RT, and drain slides.
   f. Wash slides in 4× SSC at RT for 5 min (during this time, prepare humidity chamber) and drain.
   g. Add 200 μL 5% BSA, cover, treat for 5 min at RT (during at this time prepare 4 μL fluorescein–avidin, DCS [FITC] + 796 μL 5% BSA = 800 μL; 10 μg/mL), and drain slides.
   h. Add 100 μL FITC to each slide, cover, incubate for 1 h at 37°C (during this time, place three jars with 4× SSC and one bottle with 4× SSC in a water bath at 40°C), and drain slides.
   i. Wash slide three times at 40°C in 4× SSC for 5 min each (during this period, prepare 1600 μL in 2 mL tube [80 μL goat serum + 1520 μL 4× SSC]), and drain slides.
   j. Add 200 μL goat serum to each slide, incubate for 5 min at RT in humid chamber (during this period, prepare 800 μL: 8 μL biotinylated antiavidin [10 μg/mL] + 792 μL 4× SSC), drain slides.
   k. Add 100 μL biotinylated antiavidin, cover, incubate at 37°C for 1 h, and drain slides.
   l. Wash slide four times in 4× SSC at 40°C for 5 min and drain slides.
   m. Repeat step (g).
   n. Repeat step (h).
   o. Repeat step i (during second wash, prepare propidium iodide [PI]: 16 μL PI [1 mg/mL] + 784 μL 2× SSC).
   p. Add 100 μL PI, cover, incubate in the dark for 20 min at RT in a humid chamber, and drain slides.
   q. Rinse slides briefly in 2× SSC, drain, and wipe the back of the slides.
   r. Add 50 μL (two drops) Vectashield mounting medium for fluorescence (Vector H-1000), cover with 22 mm × 32 mm × 1 mm cover glass, and store in the dark overnight.

4.7.1.3 Observation and Photography
1. Use a Zeiss Axioskop or an Olympus Fluorescence microscope.
2. Turn on the UV power supply.
3. Use a 63× oil lens.
4. There are two levels of UV: 50% for general observation and 100% for photography using a 100× oil lens.

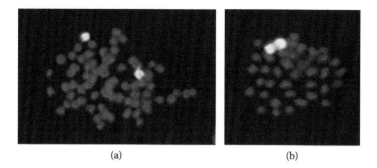

(a)   (b)

FIGURE 4.1 Fluorescence *in situ* hybridization (FISH) using ribosomal DNA labeling for identification of *Glycine tomentella* (2*n* = 78) (a) and G. max (2*n* = 40) (b). Both species are showing a pair of satellite chromosomes (bright yellow signals). (From Singh, R.J., et al. *Theor. Appl. Genet.*, 103, 212–218, 2001.)

5. Take photographs with the camera attached to the microscope and a computer.
6. Figure 4.1 shows two cells processed through the above protocol. Both cells are showing two signals; the left cell (a) is for 78-chromosome *Glycine tomentella* and right cell (b) is for *G. max* with 2*n* = 40 chromosomes.

4.8 PREPARATION OF FISH SLIDES FOR WHEAT CHROMOSOMES

This protocol is from Endo (2011).

4.8.1 CHROMOSOME PREPARATION

- *Fixative*: One part glacial acetic acid, three parts ethanol (95%–99%). Store at RT; this need not be freshly prepared (contracts with the old concept).
- *Acetocarmine stain*: Dissolve 1 g carmine powder (Merck) in 100 mL 45% acetic acid and boil for 24 h, using a reflux condenser to prevent the solution from being boiled dry. Transfer to a bottle without filtration and store at RT. Use the clear layer on the top.
- Place wheat root tips in small vials filled with distilled water and immerse in ice water for 16–20 h to collect metaphase cells.
- Fix the root tips in the fixative for one day, stain them in the acetocarmine solution for 12 h, return to the original fixative, and store for three to six days for FISH. Perform all procedures at RT. The fixed and stained root tips can be stored at −20°C until use.
- Stain the fixed and stained root tips again in the acetocarmine solution for 10–20 min.
- Make a chromosome preparation by the squash method from the stained root tips and immediately store them at −70°C or below until use.

- Remove the cover glass quickly from the frozen slide, using a razor blade, immerse the slide in 45% acetic acid at 40°C–45°C for 2–3 min, and air dry the slide at RT. The air-dried slide can be used immediately or stored in an airtight container at –20°C for several months.

4.8.2 PROCEDURE OF FISH/GISH

- Immerse the air-dried slide in the 0.15N NaOH/ethanol solution for 5 min at RT.
- Transfer the wet slide into a series of two 70% and one 99% ethanol each for 3 min at RT.
- Dry the slide quickly with a puffer.
- Apply the denatured hybridization mixture (10 μL per slide) onto the slide, place a cover glass on it and incubate the slide in a moistened chamber for 6–24 h at 30°C. (At the lower temperature (30°C) for hybridization, GISH without unlabeled blocking DNA generates as satisfactory signals as clear those generated by the GISH protocols using 37°C for hybridization).
- Remove the cover glass with a pair of forceps (when the cover glass is firmly stuck on the slide, do not apply force to remove it; instead dip the slide in 2× SSC and let the cover glass fall off) and immerse the slide for 3 min in 2× SSC at RT.
- Wash the slide briefly with distilled water and blow off water using a puffer.
- Apply the detection mixture (10 μL per slide), place a cover glass on the slide, and incubate in a wet chamber for about 1 h at 30°C.
- Remove the cover glass with a pair of forceps (when the cover glass is firmly stuck on the slide, do not apply force to remove it, but dip the slide in 2× SSC and let the cover glass fall off) and wash the slide briefly with distilled water and blow off water using a puffer.
- Apply the counter staining solution (5 μL per slide) and cover with a cover glass for fluorescence microscopic observation.
- Figure 4.2a shows Giemsa stained wheat barley (6H) addition line, while Figure 4.2b shows a FISH/GISH cell of Figure 4.2a. Nucleolar organizer regions are showing green signals.

4.8.2.1 Preparation of Stock Solution for FISH and GISH

1. *0.15N NaOH/ethanol solution*: Dissolve 6 g NaOH in 1,000 mL 40% ethanol. Store at RT.
2. *FISH probe*: Using DIG-High Prime (Roche Diagnostics) or Biotin-High Prime (Roche Diagnostics), label total genomic DNAs from the alien species and PCR-amplified subtelomeric repeat sequences of alien chromosomes, such as pSc200 and HvT01. Store the labeled probes at 4°C. FISH using total genomic DNA probes is called GISH (genomic *in situ* hybridization).
3. In situ *solution*: Prepare solution with 50% (v/v) formamide and 5% (w/v) dextran sulfate in 2× SSC. Store at 4°C.

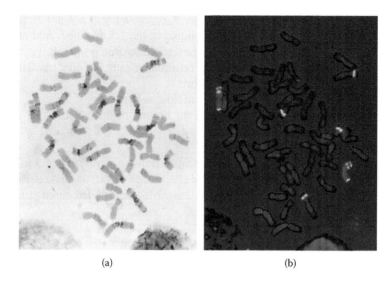

(a)   (b)

FIGURE 4.2 Disomic addition lines (6H) of wheat and barley. A cell stained by N-banding (a) and genomic *in situ* hybridization showing one pair of 6H barley chromosomes (b). (Courtesy of Endo, T.R., *Plant Chromosome Engineering: Methods and Protocols, Methods in Molecular Biology,* Springer Science + Business media, LLC., pp. 247–257, 2011.)

4. *Hybridization mixture*: Mix the labeled probe (1 part) and ISH solution (19 parts) (The proportion of the labeled probe to ISH solution should be kept minimal to attain the weakest FISH/GISH background signals, but FISH/GISH signals strong enough for observation and photography).

5. *Detection mixture*: Antidigoxigenin–fluorescein, Fab Fragments (Roche Diagnostics) or DIG-labeled probe and Streptavidin-CY3 (Invitrogen) for biotin-labeled probe diluted as recommended by the manufacturer. (The detection mixture can be diluted much more than the manufacturer recommends.)

6. *Counter staining solution*: DAPI (4', 6-diamidino-2-phenylindole) (Roche Diagnostics) diluted in an antifade solution as recommended by the manufacturer.

4.8.2.2 Purification of Labeled Probe DNA

1. Spin-column method
   a. Prepare medium by adding 10 g of Sephadex G-50 (Sigma) to 200 mL of TE (10 mM Tris-HCl, pH 8.0; ethylenediaminetetraacetic acid [EDTA]) and autoclave.
   b. Plug the bottom of a 1 mL tuberculin syringe with sterilized glass wool.
   c. Completely fill the syringe with Sephadex G-50 in TE.
   d. Remove the cap of a 1.5 mL microcentrifuge tube. Place the tube in a 15 mL plastic tube so that the tip of the syringe fits into the opening of the 1.5 mL tube.

e. Spin the tube at 1400 rpm in a bench top centrifuge for 4 min to pack down the Sephadex. Discard the liquid in the 1.5 mL tube. Add more Sephadex and recentrifuge until the packed column is 0.9 mL.

f. Replace the 1.5 mL tube with a fresh tube. Apply the nick translation reaction onto the top of the resin bed.

g. Recentrifuge at the same speed and time as before.

h. Transfer the collected effluent at the bottom of the 1.5 mL tube to a fresh tube.

2. Ethanol precipitation method

a. Add the following to 50 µL of the nick translation reaction: 6.3 µL of 4 M ammonium acetate, 140 µL of prechilled (−20°C) absolute ethanol, 5 µL of *Escherichia coli* tRNA (10 mg/mL), 5 µL of denatured salmon sperm DNA (10 mg/mL).

b. Mix well with a micropipette.

c. Leave for at least 30 min at −8.0°C or overnight at −20°C.

d. Centrifuge at 15,000 rpm for 5 min and remove the supernatant completely with a micropipette.

e. Dissolve the pellet in 50 µL of TE buffer or in 50 µL of 100% formamide.

3. Random primer DNA labeling system

This protocol has been taken from Life Technologies (https://tools.thermofisher.com/content/sfs/manuals/18187013.pdf).

a. Denature 25 ng of DNA dissolved in 5–20 µL of distilled water on a microcentrifuge tube by heating for 5 min in a boiling water bath, then immediately cool on ice.

b. Perform the following additions on ice:

i.   2 µL dATP solution

ii. 2 µL dGTP solution

iii. 2 µL dTTP solution

iv. 15 µL random primers buffer mixture

v.   5 µL (approximately 50 μC_i) [α-^{32}P] dCTP, 3000 Ci/mmol, 10 $\mu C_i/\mu L$

vi. Distilled water to a total volume of 49 µL and ix briefly.

c. Add 1 µL Klenow fragment, mix gently but thoroughly, and centrifuge briefly.

d. Incubate at 25°C for 1 h (see note 4) and add 5 µL stop buffer.

e. Dilute a 2 µL aliquot of the mixture with 498 µL of distilled water (provided by the user's laboratory) or with 498 µL of TE buffer (10 mM Tris-HCl, pH 8, 1 mM Na2EDTA).

f. Spot a 5 µL aliquot of this dilution on a glass fiber filter disk (Whatman® GF/C or equivalent). Wash the filter disk three times with 50 mL of ice-cold 10% (w/v) TCA containing 1% (w/v) of sodium pyrophosphate and once with 50 mL of 95% ethanol at RT.

g. Dry the filter under a heat lamp and determine the precipitable radioactivity by liquid scintillation counting. A result of X cpm in this test corresponds to a total of 2750× cpm incorporated in the entire incubation mixture.

For confirmation of total radioactivity in the incubation mixture, a second 5 μL aliquot of the above dilution (step vi) can be spotted on another glass fiber filter disk, dried under the heat lamp (without intervening TCA wash), and counted in a liquid scintillation counter. A result of X cpm in this test corresponds to 2750× cpm of total radioactivity in the entire reaction mixture.

Notes:

- The random primers DNA labeling system contains the deoxynucleoside triphosphates in four separate solutions. This allows the system to be used with any of the four [α-32P]-labeled nucleoside triphosphates by suitably modifying the above protocol, which is given for the case of [α-32P]-labeled dCTP. The user may also elect to mix three of the nucleoside triphosphates in a cocktail in a 1:1:1 ratio, if a particular protocol is to be consistently used.
- The random primers DNA labeling system contains a qualified control DNA that may be used to monitor the performance of the system. The incorporation curve shown on the other side was obtained with this control DNA.
- DNA probes labeled by this procedure can be used in blot hybridizations without removing the unincorporated nucleoside triphosphates (2). If desired, the labeled probe can be separated from unincorporated nucleotides by repeated ethanol precipitation (add ½ volume of 7.5 M ammonium acetate and 2 volumes of ethanol, repeat once) or using the CONCERT™ rapid PCR purification system (Cat. series 11458).
- Incubation times longer than 1 h may give higher specific activity.
- For optimal results, supercoiled DNA should be linearized or alkali denatured prior to random primer labeling:

Effect of substrate DNA configuration/denaturation on labeling:

- Substrate   Sp.Act.
- Linear pUC 19 (denatured by boiling) 2.06×10^9 cpm/μg
- Supercoiled pUC 19 (denatured by boiling) 0.12×10^9 cpm/μg
- Supercoiled pUC 19 (alkali denatured) 1.30×10^9 cpm/μg

4.8.2.3   Random Primed DNA Labeling with Digoxigenin-dUTP

1. This protocol has been taken from Fukui and Nakamura (1996).
2. Denature the linearized DNA by heating in a boiling water bath for 10 min and quickly chilling on ice.
3. Add the following to a microcentrifuge tube on ice: x μL of freshly denatured DNA, 2 μL of hexanucleotide mixture, and 2 μL of dNTP labeling mixture. Make up to 19 μL with sterile redistilled water.

   Note: 10× concentrated dNTP labeling mixture contains 1 mM dATP, 1 mM dCTP, 1 mM dGTP, 0.65 mM dTTP, and 0.35 mM digoxigenin-dUTP

4. Add 1 μL of Klenow enzyme and centrifuge briefly.
5. Incubate at 37°C for at least 1 h, to 20 h.
6. Stop the reaction by adding 2 μL of EDTA solution (0.2 M, pH 8.0) and or by heating at 65°C for 10 min.

Note: When using labeled DNA for *in situ* hybridization, removal of unincorporated dNTPs is unnecessary.

4.8.2.4. Random Primed DIG Labeling of DNA Fragment

This protocol has been taken from Roche (2006):
1. Prepare the DNA/random primer mixture as follows: 2 μL of DNA (e.g., PCR product, 25–50 ng), 1 μL of random primer (50 ng), and 9 μL of sterile water to final volume of 12 μL.
2. Incubate the DNA mixture at 95°C for 5 min, then chill on ice (a dry block heater is a safe and convenient method of heating the sample, compared with a boiling water bath, which can result in contamination of the work area).
3. Add the following reaction mixture to the denatured DNA/primer solution:
   a. 2 μL 10× DIG labeling mix. The materials for nonradioactive labeling are similar to radioactive labeling, except for the substation of [α-^{32}P] dCTP with DIG-11-dUTP.
   b. Components can be bought separately or a DIG labeling kit can be purchased from commercial suppliers [Roche Applied Science, Basel, Switzerland]. 10× DIG labeling mix (1 mM dATP, 1 mM dCTP, 1 mM dGTP, 0.6 mM dTTP, and 0.35 mM DIG-11-dUTP), 2 μL 10× Klenow enzyme buffer, 3 μL of sterile water, and 1 μL of DNA polymerase I Klenow fragment (5U/μL).
4. Incubate the final reaction solution (20 μL) at 37°C for 1 h or 20 h. Time course studies by Roche Applied Science indicate that higher yields of DIG-labeled DNA are obtained with longer incubation time (approximately fivefold increase in yield).
5. The labeled DNA fragment can then be used in hybridization techniques. It may be necessary to optimize the concentration of DIG-labeled DNA in the hybridization solution. If the concentration of the probe is too high, then background problems may appear, whereas low concentrations may result in weak signals.

4.8.3   Immunofluorescence of CENH3 of Wheat Centromere

Koo et al. (2015) applied immunofluorescence of CENH3 for examining the structure and stability of telocentric chromosomes of wheat. The protocol is given as follows:

4.8.3.1   Plant Materials and Chromosome Preparation

1. Germinate seeds in Petri plates on moist filter paper.
2. Pretreat root tips (1–2 cm) overnight in ice water.
3. Fix root tips in 3:1 ethanol:glacial acetic acid overnight.

4. Prepare slides by squash method in a drop of 45% acetic acid.

5. For immunofluorescence of CENH3, fix ice-cold treated root tips immediately using 4% paraformaldehyde in PHEMS (60 mM Pipes, 25 mM Hepes, 10 mM EGTA, 2 mM $MgCl_2$, and 0.3 mM Sorbitol, pH 6.9) for 40 min.

6. Wash with 1× PBS (10 mM sodium phosphate and 140 mM NaCl, pH 7.0) for 20 min on ice.

7. Treat root tips with 2% cellulase, 1% pectinase (Sigma, St. Louis, MO), and 1% pectolyase in PHEM at 37°C for 1 h.

8. Prepare slide by squash method in a drop of water on a poly-L-lysine-coated slide (Sigma).

9. Store all preparations at −70°C until use.

4.8.3.2 Immunodetection of CENH3 and FISH

1. Synthesize peptide antigen, RTKHPAVRKTKALPKK, and use to immunize rabbits at Thermo Fisher Scientific (www.thermofisher.com).

2. Purify the raised antisera using an affinity sepharose column comprising the aforementioned peptide.

3. Check specificity of the antibody by immunostaining of root tips and pollen mother cells of wheat.

4. Dilute the rabbit anti-CENH3 to 1:1000 in TNB buffer [0.1 M Tris-HCl, pH 7.5, 0.15 M NaCl, and 0.5% blocking reagent (Roche, Indianapolis, IN)].

5. Add approximately 100 μL of the diluted antibodies to each slide and incubate the slides in a humid chamber at 37°C for 2–3 h.

6. Wash in 1× PBS three times.

7. Add 100 μL rhodamine-conjugated goat antirabbit antibody to each slide (Jackson ImmunoResearch, West Crove, PA) (1:100 in TNB buffer).

8. Incubation and washes are the same as for the primary antibody.

9. Label DNA probes of the CRWs, pAs1, pSc119, and the other single-gene probes with digoxigenin-11-dUTP, biotin-16-dUTP, and/or DNP-11-dUTP (PerkinElmer, Waltham, MA), depending on whether two or three probes are used in the FISH experiment.

10. After posthybridization washes, detect the probes with Alexafluor 488 streptavidin (Invitrogen, Carlsbad, CA) for biotin-labeled probes and with rhodamine-conjugated antidigoxigenin for dig-labeled probe.

11. Detect the DNP-labeled probe with rabbit anti-DNP, followed by amplification with a chicken antirabbit Alexa Fluor 647 antibody (Invitrogen).

12. Multicolor immune-FISH detection:

13. Counterstain chromosomes post multicolor immune-FISH detection with 4′, 6-diamidino-2-phenylindole (DAPI) in Vectashield antifade solution (Vector Laboratories, Burlingame, CA).

14. Capture images with Zeiss Axioplan 2 microscope (Carl Zeiss Microscopy LLC, Thornwood, NY) using a cooled CCD camera Cool-SNAP HQ2 (Photometrics, Tucson, AZ) and Axio Vision 4.8 software.

15. Process the final contrast of the images using Adobe Photoshop CS5 software.

4.8.3.3   Sequential detection of CENH3, CRWs, pSc119, and pAs1

1. Incubate slides with anti-CENH3 overnight at 4°C in a wet chamber.
2. Washing in 1× PBS, incubate slides with the appropriate secondary antibody at 37°C for 50 min.
3. Refix slides with 4% paraformaldehyde at RT for 30 min.
4. Denature slides in 70% formamide in 2× SSC, 80°C for 2 min.
5. Wash slides in ice-cold 1× PBS for 5 min.
6. Apply DNA probe, CRWs to the slides.
7. Posthybridization washes and signal detection are the same as for FISH procedure.
8. After recording the CENH3 and CRWs signals, remove cover slips and wash slides in 4T buffer (4× SSC/0.05% Tween 20) for 1 h at 37°C and refix with 4% paraformaldehyde and dehydrate in an ethanol series.
9. Reprobe slides with pAs1 and pSc119 to detect additional sequences on the same chromosome.

4.8.3.4   Genome-specific markers and PCR

1. Perform PCR with 15 μL of the reaction mixture containing 1× PCR buffer (Bioline USA Inc., Taunton, MA), 2 mM MgCl$_2$, 0.25 mM dNTPs, 5 pmol forward primer, and reverse primer, respectively, 0.02 U/μL *Taq*-DNA polymerase (Bioline), and 20 ng genomic DNA.
2. Resolve PCR products on 2.5% agarose gels and visualize by ethidium bromide staining under UV light.
3. Figure 4.3a through h shows identification of telocentric chromosomes using above procedure. On the contrary, Figure 4.4 shows identification of Chinese Spring wheat, Ditelo for the short arm and long arm of D-genome using CENH3, CRWs, and pAs1 techniques.

4.9   GENOMIC *IN SITU* HYBRIDIZATION

Genomic *in situ* hybridization (GISH) is a powerful cytological method to distinguish parental chromosomes or chromosome segments in an interspecific or intergeneric hybrids. It has been proved to be a boon to cytogeneticists in identifying precisely the inserted region in the recipient parent from the alien species and in examining the evolutionary relationship of crops.

- The protocol for GISH essentially is the same as for the FISH except for the blocking genomic DNA. The GISH protocol is defined to eliminate most of the cross-hybridization between total genomic DNA from the two species. Optimum results are obtained when blocking DNA exceed the concentration of probe DNA by one hundred fold. Blocking genomic DNA fragments of 100–200 bp length are obtained by autoclaving the total genomic DNA from each parental species.
- For GISH mixture, the yield of labeled probe can be increased by use of carrier DNA. The role of block and carrier DNA is combined in addition of unlabeled total genomic blocking DNA at 35× the concentration of probe DNA

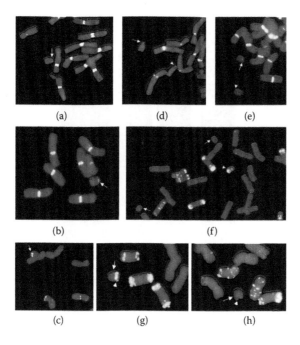

FIGURE 4.3 Localization of CENH3 (arrow in a), CRWs (arrow in b) and pAs1 (arrow in c) on Dt6DS; localization of CENH3 (d), CRWs (e), and pAs1 (f) on dDt6D; arrows and arrowhead indicate the 6DS and 6DL telosomes, respectively. Two-color detection of single-gene probe, 6S-2 and pAs1 on Dt6DS (g) and dDt6DS (h); the hybridization signal for single copy probe 6S-2 (red dots) is indicated by arrows and the pAs1 was labeled with green colors by arrowheads. (From Koo, D.H., et al. 2015. *PLoS ONE|* DOI:10.1371/journal.pone.0137747.

immediately after stopping the NT reaction with EDTA. The BIONICK Labeling System is used for biotin-labeled DNA probes by nick translation. (BIONICK™ Labeling System [Cat. no. 18247-015; Lot no. KKF712]).

- In this method, the genome of the second parent (unlabeled) is used as a blocking DNA to avoid nonspecific hybridization due to similarity of the two parental genomes. Therefore, both parental genomes (the probe and the blocking DNA) must be used together in the same hybridization mixture. The proportion of probe:blocking DNA is determined to avoid the detection of the second parent (Brammer et al. 2013).

4.9.1 DNA Isolation

In GISH, isolation the genomic DNA of the probe and blocking DNA should be free from contamination. Usually, leaves are most commonly used for DNA extraction.

4.9.1.1 CTAB Method
1. Weigh approximately 300 mg of leaf tissues and put in a 1.5 mL centrifuge tube.
2. Macerate carefully in liquid nitrogen avoiding defrosting the tissue.

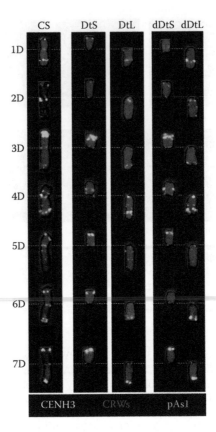

FIGURE 4.4 Immuno-FISH-based karyotype of D-genome chromosomes of wheat and their derived telosomes using CENH3 (white), CRWs (red), and pAs1 (green) as probes. CRWs (red signals) colocalized with CENH3 (white signals) in most of the chromosome except dDt1DS, 4D, Dt4DS, dDt4DS, Dt5DL, and Dt5DL. The centromeric regions of chromosome or chromosome arm were seen as pinkish red colors because the CRWs (red signals) are abundant in the centromeric region and much brighter than CENH3 signals except in the above-mentioned telosomes. The dDt1DS stock contained multiple chromosome rearrangement including inversion, deletion, and centromere shift. Note that the CRWs were not detected in Dt4DS and dDt4DS; instead, the pAs1 signal was overlapped with the CENH3 signal in these telosomes. A very faint pAs1 FISH site was detected in the terminal region of dDt6DS, indicating a terminal deletion. Short arm and long arm telosomes present in the ditelosomic stocks are represented as (DtS) and (DtL), respectively, and short arm and long arm telosomes present in the double ditelosomic stocks are represented as (dDtS) and (dDtL), respectively. (From Koo, D. H., et al., *PLoS ONE*, 2015, doi:10.1371/journal.pone.0137747.)

3. Add 700 μL of preheated (65°C) isolated buffer (2% CTAB; 100 mM Tris-HCl, pH 8.0; 20 mM EDTA, pH 8.0; and 1.5 M NaCl) and mix well.
4. Incubate the samples for 60 min at 65°C in water bath and mix gently every 10 min.
5. Remove from the water bath and let cool to RT (~24°C) for 5 min.

6. Add 700 μL of CIAA (chloroform: isoamyl alcohol, 24:1 v/v) and mix gently for 10 min.
7. Centrifuge for 7 min (10,000 rpm at RT).
8. Transfer the supernatant to new centrifuge tubes, add again 700 μL CIAA, and mix gently for 10 min.
9. Centrifuge for 7 min (10,000 rpm at RT).
10. Transfer the supernatant to new centrifuge tube and add 500 μL of cold (−20°C) isopropanol. Mix gently to precipitate the DNA and incubate for at least 30 min at −20°C.
11. Centrifuge for 5 min (10,000 rpm at RT).
12. Discard the supernatant carefully in order to not lose the pellet.
13. Wash the pellet with 600 μL of cold 70% ethanol and discard the 70% ethanol.
14. Wash the pellet with 600 μL of cold 96% ethanol, discard the 96% ethanol, and dry the pellet at RT.
15. Resuspend the pellet in 100 μL of 10 mM Tris-HCl (pH 8.0) or ultrapure distilled water.
16. Add 3 μL of 10 mg/mL RNase A, mix, and incubate for 1 h at 37°C.
17. Store samples at −20°C or −80°C. For long time storage, the best results are obtained when pelleted materials are stored in 70% ethanol.

4.9.1.2 Selective Precipitation of Polysaccharides

1. Add 500 μL of the precipitation solution (10 mM Tris-HCl [pH 8.0] and 250 mM NaCl).
2. Dissolve the pellet by vortexing. The complete dissolution of the pellet is important to not to lose DNA. Samples with much polysaccharide contamination tend to dissolve more slowly.
3. Add 180 μL of cold 100% ethanol, mix the solution by vortexing, and put immediately in chopped ice.
4. Put in the refrigerator (10°C) for 20 min or in a freezer (−20°C) overnight.
5. Centrifuge for 20 min (10,400 rpm at 4°C).
6. Transfer the aqueous phase to a new tube. In this step, the DNA is in the aqueous phase and the pellet may be discarded.
7. Add 700 μL of isopropanol and mix gently inverting the tubes approximately 50 times.
8. Centrifuge for 20 min (10,400 rpm at 4°C).
9. Discard the supernatant and dry the pellet at RT.
10. Add 500 μL of 70% cold ethanol and invert the tube approximately 20 times.
11. Centrifuge for 20 min (10,400 rpm at 4°C).
12. Discard the supernatant and dry the pellet at RT.
13. Resuspend the pellet in 10 mM Tris-HCl (pH 8.0) or ultrapure distilled water.

4.9.1.3 DNA Quantification

1. Quantify DNA prior to probe labeling and preparation of blocking DNA.
2. Preform an electrophoresis in 0.8% agarose gel with an aliquot of each isolated DNA.

3. Use ʎ-DNA as a reference with different amounts (e.g. 50, 100, and 150 ng).
4. Compare between reference bands and bands of the isolated DNA.
5. Prepare an aliquot containing 5–50 μg of the previously quantified DNA in 100 μL in a 1.5 μL Eppendorf tube (diluted in ultrapure distilled water).
6. Keep the 1.5 μL Eppendorf tube in a closed flask to avoid both the opening of the Eppendorf tube and the direct contact of the sample with the autoclave steam. Keep the flask in the autoclave.
7. Turn on the autoclave and when the temperature reaches 121°C, mark 5 min, and then turn off.
   After removing the Eppendorf tube from autoclave, cool and spin down the volume. Run an electrophoresis with the autoclaved DNA and 190 bp ladder (as reference) in a 0.8% agarose gel. The fragmented DNA must be between 100 and 300 bp.
   For GISH in wheat–rye hybrids, the blocking DNA (wheat DNA) must be at the concentration of 500 ng/μL, due to the concentration of probe of 50 ng/μL (ratio of 1:10, probe:blocking DNA). However, for hybrids between other species, the concentration of blocking DNA may be higher, if the proportion probe:blocking DNA is different. For example, if the proportion to be used is 1:20, the blocking DNA should be at 1 μg/μL.
8. Add 2 volume of cold absolute ethanol and 0.1 volume of 3 M sodium acetate (or 0.05 volume of 7.5 M sodium acetate) to precipitate the DNA.
9. Mix gently by inverting and store overnight at −20°C.
10. Centrifuge for 20 min (14,000 rpm at RT).
11. Wash the pellet with 1 mL of 70% ethanol.
12. Centrifuge for 5 min (14,000 rpm at RT).
13. Dry the pellet at RT or at 37°C.
14. Resuspend the pellet in 10 mM Tris-HCl (pH 8.0) or in ultrapure distilled water, in order to reach the required concentration (500 ng/μL). Take into account that there are losses in the total quantity of DNA during the steps of precipitation and resuspension.

4.9.1.4 Nick Translation

The nick translation reaction is usually performed with commercial labeling kits, which should be performed according to the manufacturer's recommendations. However, the following protocols are followed:

1. Prepare the nick translation mixture in a centrifuge Eppendorf tube surrounded by chopped ice without enzymatic solution.
2. Vortex the mixture, spin down the volume, and add the enzymatic solution rapidly.
3. Mix gently, spin down the volume, and put the Eppendorf tube either in the thermocycler or in a thermos block at the recommended temperature. To find out if the reaction time recommended by the manufacturer was

sufficient for obtaining fragments with 200–300 bp, the labeling process should be temporarily suspended by maintaining the Eppendorf tube in chopped ice. Meanwhile, an electrophoresis in 0.8% agarose gel should be performed with an aliquot of the reaction. If the DNA is sufficiently fragmented, add the stop buffer. Otherwise, the reaction must continue as long as necessary for correct DNA fragmentation.

4. After adding the stop buffer, add 2 volume of cold absolute ethanol and 0.1 volume of 3 M sodium acetate, in order to precipitate DNA.
5. Mix gently by inverting and put in the freezer overnight.
6. Centrifuge for 20 min (14,000 rpm at RT), discard the supernatant, and dry the pellet at RT or at 37°C.
7. Resuspend the pellet in 15–20 µL of 10 mM Tris-HCl (pH 8.0) and store at −20°C.

4.9.2 COLLECTION OF ROOTS

Germination of seeds, collection of roots, pretreatment, fixation, and preparation of chromosome spread have been described before.

4.9.3 GENOMIC *IN SITU* HYBRIDIZATION

Treatment of slides: Examine the slides and identify the cells with well-spread chromosomes. Slides stored for a long time must be dipped in Carnoy's solution for 15 min followed by an alcoholic series of 70% and 100% ethanol for 5 min each. The Carnoy's solution helps to better fix the chromosome structure.

- Dry the slides for 30 min at 50°C–60°C. This drying is important because the grip of the chromosomes to the blade is improved during the process.
- Cool the slides for 5–10 min at RT.
- Add 50 µL of 100 µg/mL RNase A (10 mg/mL RNase A solution diluted in 2× SSC [300 mM NaCl and 30 mM $Na_3C_6H_5O_7.2H_2O$] at the proportion of 1:100), cover with a plastic coverslip (made from Parafilm) and incubate in a moisture chamber for 1 h at 37°C.
- Wash slides three times in 2× SSC for 5 min each. After each wash cycle, the used 2× SSC must be replaced by fresh 2× SSC.
- Add 50 µL of 10 mM HCl, cover with plastic coverslip, and keep for 5 min.
- Add 50 µL 10 µg/mL pepsin (1 mg\mL pepsin diluted in 10 mM HCl at the proportion of 1.5:100), cover with plastic coverslip, and incubate at 37°C for 20 min. Remove excess HCl with absorbing paper before adding the pepsin.
- Wash the slides three times in 2× SSC for 5 min each.
- Treat the chromosome preparation in 4% paraformaldehyde for 10 min. Paraformaldehyde is highly toxic and carcinogenic.
- Wash slides three times in 2× SSC for 5 min each.
- Dehydrate slides in 70% and 100% ethanol for 3 min each.
- Air dry the slides for 1 h at RT.

4.9.4   PREPARATION OF GISH HYBRIDIZATION MIXTURE

- 50% formamide
- 10% dextran sulfate
- 2× SSC (2.5–5 ng/μL of probe and 25–50 ng/μL blocking with final volume of 10 μL/slide). In case of using directly labeled probes, the mixture preparation and all the following steps must be done in partial darkness (avoid direct incidence of light). The formamide destabilizes the DNA molecule, helping in the denaturation.
- For probe denaturation, incubate the mixture at 75°C for 10 min in a water bath. Immediately, put the mixture on ice for at least 5 min to keep the two DNA strands open.
- Spin down the mixture and then add 10 μL per slide, cover with an 18 mm × 18 mm glass cover slip and denature the slide at 73°C for 19 min.
- Seal the cover slip with rubber glue and incubate the slide in a moist chamber at 37°C overnight (~16 h) or up to a day and a half.

Components	Quantity	Final concentration
100% formamide	5 μL	50%
50% dextran sulfate	2 μL	10%
20× SSC	1 μL	2%
Probe	0.5–1 ng/μL	~2.5–5 ng/μL
Blocking DNA	0.5–1 ng/μL	~25–59 ng/μL
Ultrapure distilled water	E μL	

4.9.5   POSTHYBRIDIZATION

In this step, flasks with SSC solutions and the Coplin jar (with the solution of the first wash) must be previously in the water bath at 42°C. Before checking the temperature of the water bath, the temperature inside the Coplin jar must also be checked. The same jar can be used for all the baths by discarding the anterior SSC solution and adding the next one. Furthermore, the function of the wash procedures is to remove the excess of materials from the *in situ* hybridization, mainly nonhybridized probes and incorrectly hybridized ones.

- Remove the rubber glue with tweezers without moving the cover slip.
- Dip the slides with coverslips in the prewarmed Coplin jar. After dipping the last slide remove the coverslips carefully to avoid damage to the chromosomes. The time count begins only after the last coverslip is removed.
- Wash slides twice in 2× SSC for 5 min each at 42°C.
- Wash the slides twice in 0.1× SSC (15 mM NaCl and 1.5 m M $Na_3C_6H_5O_7.2H_2O$) for 5 min (stringency of 73%) at 42°C.
- Wash slides twice in 2× SSC for 5 min each at 42°C. In the second wash, remove the Coplin jar from the water bath.

- Wash slides once in 2× SSC for 5 min at RT.
- Wash the slides once in 4× SSC + 0.1% Tween 20 (600 mM NaCl and 60 mM $Na_3C_6H_5O_7.2H_2O$ and 0.1% Tween 20) at RT.
- Note: During the washes of slides with directly labeled probes, the procedures must also be done in partial darkness due to the presence of fluorochromes. Moreover, GISH procedures end after these washes when using this type of probe and the results may be already visualized in an epifluorescence microscope after adding DAPI-Vectashield as explained below:
- Dry the excess of SSC, add the secondary antibody mixture (antibody and 1% BSA in 4× SSC + 0.1% Tween 20, according to the manufacturer's recommendation), cover with a plastic coverslip and incubate at 37°C for 1≈h in a dark moist chamber.
- Wash the slides three times for 10 min each in 4× SSC + 1% Tween 20 at 42°C. These washes are needed for removing the excess of antibodies.
- Dry the excess SSC, add 8 μL of DAPI (4',6 diamidino-2-phenylindole; 2 μg/mL) and Vectashield antifade (1:1, v/v) and mount the slide with a 22 mm × 22 mm glass coverslip. Seal coverslip with colorless nail polish and allow it to dry for at least 1 h in the dark.
- Analyze the slides in an epifluorescence microscope with the adequate fluorescence filter.

4.10 MULTICOLOR GENOMIC *IN SITU* HYBRIDIZATION (McGISH)

Multicolor GISH is an excellent cytogenetic tool for simultaneous discrimination of each genome and identification of diploid progenitors of allopolyploids, simultaneous mapping of different DNA sequences, physical ordering of multiple probes in a single chromosome, genome allocation of the gene of interest, detection of chromosomal aberrations, and examining chromosome organization in interphase nuclei (Mukai 1996). Shishido et al. (1998) developed multicolor genomic *in situ* hybridization to identify somatic hybrids between *Oryza sativa* cv. Kitaake (AA, 2n = 24) and *O. punctata* (BBCC, 2n = 48) and the progeny rescued from embryo culture. The procedure is as follows:

1. Label total DNA of diploid rice species, *O. sativa* (AA), *O. punctata* (BB), and *O. officinalis* (CC) with biotin-16-dUTP (Boehringer Mannheim) or digoxigenin-11-dUTP (Boehringer Mannheim) by the standard random primed labeling protocol.
2. Denature hybridization mixture (100 ng labeled probe/slide + equal parts of 50% formamide and 2× SSC) for 10 min at 90°C and cool immediately on ice (0°C).
3. Denature chromosome spreads in 50% formamide/2× SSC for 6 min at 70°C with the hybridization mixture (Fukui et al. 1994) and then hybridize for three to four days at 37°C.
4. Wash twice in 2× SSC, once in 50% formamide/2× SSC, and once in 4× SSC each for 10 min at 40°C.

5. Apply a drop of fluorescein–avidin (1% Vector laboratories, CA) + 1% bovine serum albumin (BSA) in BT buffer (0.1 M sodium hydrogen carbonate, 0.05% Tween 20, pH 8.3) on chromosome spread.
6. Incubate at 37°C for 1 h.
7. Wash slides three times (10 min each) at 37°C.
8. Amplify again (repeat step 5).
9. Apply a drop on chromosome spread antidigoxigenin-rhodamine in 5% goat biotinylated antiavidin and antirhodamine in 5% goat serum in BT buffer, incubate at 37°C for 1 h.
10. Wash with BT buffer three times at 37°C (5 min each).
11. Apply a drop of 1% fluorescein-avidin and antisheep-Texas Red (1% Vector) in 1% BSA in BT buffer to chromosome spread.
12. Incubate at 37°C for 1 h.
13. Wash twice with BT buffer, and once with 2× SSC at 40°C (10 min each).
14. Counterstain chromosomes with 1 μg/mL 4′,6-diamidino-2 phenylindole (DAPI) in an antifadant solution (Vector Shield, Vector).
15. Figures 4.5 and 4.6 shows multicolor genomic staining of rice species chromosomes.

Note: Block chromosome spread with 5% bovine serum or goat serum albumin in BT buffer at 37°C for 5 min before each immunocytochemical step.

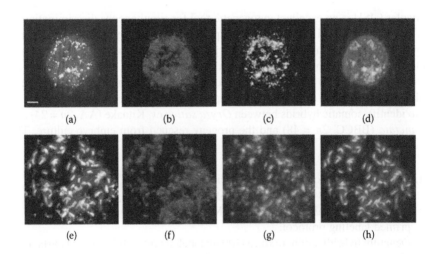

FIGURE 4.5   (a–h) McGISH of PU-2 with the two genomic probes of the A (*red*) and C (*green*) genomes. Counterstaining was applied with DAPI (*blue*). (a and e) Discrimination of the three deferent genomes in a nucleus and a chromosome complement. (b and f) Distribution of the A genome within the nucleus and detection of the A genome chromosomes with *red fluorescence*. (c and g) Distribution of the C genome within the nucleus and detection of the C genome chromosomes with *green fluorescence*. (d and h) DAPI-stained nucleus and chromosomes. *Bar*: 5 μm. (From Shishido, R., et al., *Theor. Appl. Genet.*, 97, 1013–1018, 1998.)

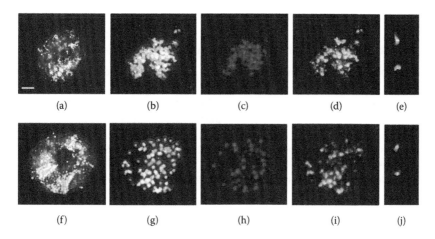

(a)   (b)   (c)   (d)   (e)

(f)   (g)   (h)   (i)   (j)

FIGURE 4.6 (a–j) McGISH of PU-289 with the different combination of the genomic probes, the A and B (*upper panels*) or the A and C (*lower panels*) genomes, respectively. (a) Discrimination of the three different genomes with *red* (A genome) and *green* (B genome) *fluorescence* in the nucleus. The nucleus was counter-stained with DAPI (*blue*). (b) Discrimination of the chromosomes belonging to the three different genomes with *red* (A genome) and *green* (B genome) *fluorescence*. (c and h) Detection of the A genome chromosomes with *red fluorescence*. (d) Detection of the B genome chromosomes with *green fluorescence*. (e) Insertion of the B genome chromosomal fragments into the C genome chromosomes detected. (f) Discrimination of the three different genomes with *red* (A genome) and *green* (C genome) *fluorescence* in the nucleus. (g) Discrimination of the chromosomes belonging to the three different genomes with *red* (A genome) and *green* (C genome) *fluorescence*. (i) Detection of the C genome chromosomes with *green fluorescence*. (j) Insertion of the A genome chromosomal fragments into the B genome chromosomes detected. *Bar*: 5 μm. (From Shishido, R., et al. *Theor. Appl. Genet.*, 97, 1013–1018, 1998.)

4.11 MULTICOLOR FISH FOR WHEAT GENOME

4.11.1 GENOMIC LIBRARY CONSTRUCTION

- Komuro et al. (2013) examined 2000 plasmid wheat clones for presence of signals after FISH. Among them, 47 clones produced strong discrete signals on wheat chromosomes (Table 4.1, Figure 4.7a through c).
- Extract genomic DNA from leaves of the wheat cultivar Chinese Spring using the standard CTAB procedure.
- Digest the genomic DNA solution (1 μg/μL) partially using DNase I (Roche Diagnostics, Mannheim, Germany).
- Collect 800–1000 bp fraction from an agarose gel.
- After an adenosine addition to the ends of digested DNA fragments using *Taq* polymerase, ligate fragments to a pGemT-easy vector (Promega, Fitchburg, Wis., USA) according to the manufacturer's instructions.
- Electroporate a bacterium strain (stbl4; Invitrogen, Carlsbad, Calif., USA) with the desalted pGemT-easy plasmid solution according to the manufacturer's instructions.

TABLE 4.1

List of 47 FISH-positive Wheat Clones Identified in this Work

Signal distribution or homologous sequence	Clone No.
All centromere positions	pTa-103, pTa-179[a], pTa-435, pTa-636, pTa-725, pTa-925, pTa-k165, pTa-k404, pTa-k439, pTa-k465, pTa-k530, pTa-k548, pTa-k556, pTa-k597, pTa-k604
Centromere related[b]	pTa-1, pTa-k18[a]
Diffuse signals enriched in the A-genome chromosomes	pTa-k229[a], pTa-k423, pTa-k630, pTa-k700
Diffuse signals enriched in the B-genome chromosomes	pTa-k532
Diffuse signals enriched in both A- and B-genome chromosomes	pTa-k187, pTa-k288[a], pTa-k354
pAs1	pTa-173[a]
pAs1 related[b]	pTa-535[a], pTa-s53
pSc119	pTa-86[a], pTa-505, pTa-835
Distinct bands on multiple chromosomes	
pTa-713 pattern[b]	pTa-551, pTa-713a, pTa-779, pTa-885
pTa-k566 pattern[b]	pTa-458, pTa-k566[a]
pTa-465 pattern[b]	pTa-465[a]
Ribosomal DNA sequence 5S	pTa-72
Ribosomal DNA sequence 28S	pTa-k374, pTa-s208, pTa-s309
Microsatellite related	pTa-272, pTa-275[a], pTa-451, pTa-s120[a], pTa-s126[a]

Source: From Komuro, S., et al. *Genome* 56:131–137, 2013.

Note: Sequences presented here are registered in the gene bank as accession numbers KC290868–KC290916.

[a] Sequences used to reveal the FISH signal distribution.

[b] Newly identified FISH-positive sequences in wheat.

- The exception was that there was no PCR prescreening, as this portion of the work was done before the work described in Kato (2011) and did not include that improvement. In the screening, only Texas red labeling was used (no reprobing).

4.11.1.1 Slide Preparation

1. Prepare slides from root tips of Chinese Spring wheat seeds at 30°C.
2. Treat root tips with 1.0 MPa nitrous oxide gas in an airtight stainless steel chamber for 90 min.
3. Fixed in 90% acetic acid for 2 min.
4. Wash the root tips immediately in a TE buffer (10 mmol/L Tris-HCl, 1 mmol/L EDTA, pH 7.6) for 5 min.
5. Cut four root meristems and put into a 1.5 mL centrifuge tube containing 10 μL pectolyase Y-23 (1%) (Kikkoman Co., Tokyo, Japan) and cellulase Onozuka R-10 (2%) (Yakult Honsha Co., Ltd. Minato-ku, Tokyo, Japan) in citric buffer (sodium citrate 5 mmol/L, EDTA 5 mmol/L, pH 5.5).

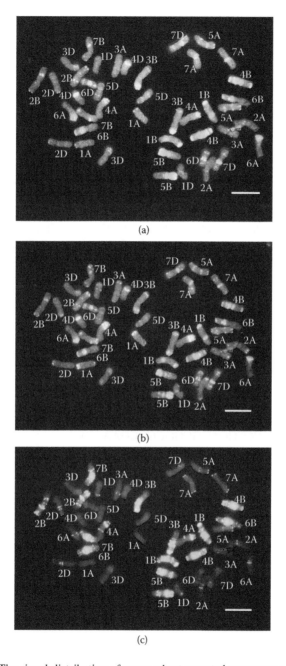

(a)

(b)

(c)

FIGURE 4.7 The signal distribution of seven wheat-repeated sequences revealed by four repeated *in situ* hybridization of the same chromosome spread. (a) The signals of pTa-535 (red), pTa-713 (green), and pTa-86 (light blue). (b) The signals of pTa-535 (red), pTa-713 (green), pTa-86 (light blue), pTa-465 (yellow), and pTa-k566 (purple). (c) The signals of pTa-173 (red), GAA microsatellite sequence (green), and pTa-86 (light blue). Scale bar = 10 μm. (From Komuro, S., et al. *Genome,* 56, 131–137, 2013.)

6. Transfer tube to a 37°C water bath, and digest root sections for 90 min.
7. After digestion, wash root sections in TE buffer.
8. After aspiration of the TE buffer, add 50 µL of 70% ethanol to the tube. Then agitate the tube several times to disperse the digested cell sections into a suspension of single cells.
9. Centrifuge tube at 5000 rpm for 1 min to collect root cells.
10. Discard 70% ethanol, and wash the cells with 100% ethanol to remove any remaining moisture.
11. Spin the cells down again and aspirate ethanol.
12. Resuspend the cells in 30 µL of 100% acetic acid.
13. Place clean glass slides in a humidity chamber, which was a small open-top cardboard box (10 cm × 10 cm × 20 cm) lined on the bottom with moistened paper towels and with two wooden rods.
14. Add four droplets (0.5 µL each) of an acetic acid cell suspension onto a clean glass slide in two lines of two droplets.
15. Allow the slides to dry slowly in the humidity chamber.
16. Fix spread cell on the slides with a 10% formaldehyde solution for 5 min by immersing the slides.
17. Wash 10% formaldehyde solution from the slides by spraying with 70% ethanol, followed by spraying with 100% ethanol.

4.11.1.2   Probe Labeling Using VentR (exo-)Polymerase

1. Isolate plasmids using the regular mini alkali lysis procedure.
2. Amplify the inserts of each plasmid by PCR using primers designed from the cloning site sequences (i.e., pGEM-T easy forward–22 5'-GACGTCGCATGCTCCCGGCCGCCA-3= and pGEM-T easy reverse+12 5'-TGGTCGACCTGCAGGCGGCCGCGAA- 3').
3. After ethanol precipitation, resuspend each PCR product (about 10 µg) TE buffer (250 ng/µL, 40 µL).
4. Add 2 µL of magnesium sulfate (100 mmol/L) to the tubes, and digest the PCR products by adding 100 mU DNase I.
5. Confirm the DNA lengths (~100 bp) by 1% agarose gel electrophoresis and ethidium bromide staining.
6. For Texas red labeling, add the following components 200 µL PCR tube:
    a.   5 µL digested DNA
    b.   thermo buffer (10×) 2 µL
    c.   Texas red-5-dATP 0.2 µL (Perkin Elmer)
    d.   0.4 µL dGTP-dCTP mix (0.2 mmol/L each)
    e.   0.2 µL VentR (exo-) DNA polymerase (New England Biolabs, Japan, Sumida-ku Tokyo, Japan)
7. Add nuclease-free water for a final volume of 10 µL.
8. For fluorescein labeling, use a 200 µL PCR tube:
    a.   Digest 5 µL DNA
    b.   2 µL thermo buffer (10×)
    c.   0.2 µL fluorescein-12-dUTP

   d. 0.4 µL dCTP (0.2 mmol/L)

   e. 0.2 µL VentR (exo-) DNA polymerase

9. Add nuclease-free water a final volume of 10 µL.

10. In a thermal cycler, subject the tubes to 100 cycles of 94°C for 10 s and 20°C for 1 min. In each cycle, the digested DNA strands anneal to each other after denaturation and at ~ 70°C DNA synthesis incorporates the labeled base analogs.

11. Coprecipitated the labeled DNA with digested salmon sperm DNA (10 µg) by the regular ethanol/sodium acetate (pH 5.2) precipitation procedure.

12. Redissolved the probe 2× SSC: 1× TE buffer (0.3 mol/L sodium chloride, 30 mmol/L sodium citrate, 1 mmol/L Tris-HCl, 10 mmol/L EDTA, pH 7.6).

13. Mix 30-mer synthetic oligos of GAA×10 and TTC×10 (100 pmol each) and label using the above-mentioned terminal extension method with fluorescein–12-dUTP and dATP-dGTP-dCTP mix (0.2 mmol/L each).

4.11.1.3 FISH and Reprobing Procedures

1. Place a slide with a good chromosome spread on a metal plate on ice.

2. Drop a probe mixture (e.g., 10 ng/µL fluorescein-labeled pTa-713 and 10 ng/µL Texas red labeled pTa-535, total volume 2 µL) on the slide.

3. Immediately apply a plastic coverslip (4 mm × 4 mm, rain X-coated).

4. Apply a clean slide and secure with paper clips to make a sandwich structure that prevents dehydration.

5. Maintain slide at 100°C for 5 min to denature the chromosomal DNA and probes.

6. Incubate the slide at 55°C in a sealed storage container with a moist paper towel for 2–3 h.

7. Wash (55°C, 2× SSC, 1 mmol/L EDTA, 20 min)

8. Allow the slides to drain.

9. Apply a drop of Vectashield mounting medium with DAPI (Vector Laboratories, Inc., Burlingame, Calif.,USA).

10. Place a 24 mm × 32 mm coverslip to over the cells.

11. Capture fluorescent signals from each of the three channels (DAPI, fluorescein, and Texas red) through a 100× plan apochromatic oil objective using a BX60 epifluorescence microscope (Olympus, Shinjuku-ku, Tokyo, Japan) with an EOS Kiss X2 camera (Canon, Oota-ku, Tokyo, Japan) and a camera adaptor provided by MeCan Imaging, Inc. (Fujimino-shi, Saitama, Japan, http://www.mecan.co.jp/index.html).

12. Capture the fluorescence of the DAPI channel using a U-MWBV filter (excitation 400–440 nm, violet light, emission 475 nm, Olympus) because the exposure to UV light (330–385 nm) using 100× oil lens damages the chromatin DNA structure and reduces FISH signal intensity significantly.

13. After capturing images, remove the coverslip and wash the slide gently with 70% ethanol.

14. Submerge the slide in a boiling 2× SSC buffer (100°C) for 5 min to remove probes.

15. Wash the slide with distilled water and then rinse with 70% ethanol briefly and air dry.
16. Examine the dried slide using a phase contrast microscope to confirm the integrity of the chromosome spread.
17. Then apply one drop of DAPI (2 µg/mL) containing distilled water and apply a coverslip to confirm the absence of remnant fluorescent signals under the epifluorescence microscope. At the same time, capture the photographs of the chromosome spread on each channel.
18. This background picture is very useful to deduct the backgrounds from the pictures with fluorescent signals.
19. Remove the coverslip by applying 70% ethanol, and dry.
20. The slide is then ready for another round of *in situ* hybridization.

The order of FISH signal detections used is as follows:

1. In the first FISH trial, apply pTa-465 (Texas red) and pTa-k566 (fluorescein) probes.
2. After removal of probes, apply pTa-535 (Texas red) and pTa-713 (fluorescein) probes.
3. Next, apply pTa-173 (Texas red) and pTa-86 (fluorescein) probes.
4. Finally, apply GAA microsatellite (Texas red) probe to the slide.
5. After capturing red and green images, subtract the background data using Photoshop software (Adobe Systems, San Jose, Calif., USA).
6. Some of the green or red color images were changed to blue, yellow, or purple on the software (pseudocolored) and overlay.

4.11.2 PRIMED *IN SITU* (PRINS) DNA LABELING

PRINS DNA labeling technique, first described by Koch et al. (1989), is an alternative to FISH for the detection of nucleic acid. This technique involves labeling chromosomes by annealing an oligonucleotide DNA primer to the denatured DNA of chromosomes spread on slide glass and extending it enzymatically *in situ* with incorporation of labeled nucleotides. The PRINS technique has been used extensively in human cytogenetics for mapping of repetitive and low-copy sequence, for chromosome identification, for detection of aneuploidy in sperm cells, the analysis of the human chromosome complement of somatic hybrids, for specifying chromosome rearrangements by combination of PRINS labeling of chromosome-specific alphoid sequences and chromosome painting (Pellestor et al. 1996; Gosden 1997).

The reliable and reproducible detection of single-copy sequence below 10 kb in plants with large genome has been difficult by FISH. Menke et al. (1998) developed a procedure of PRINS for detection of repetitive and low-copy sequences on plant chromosomes. For chromosome preparation, follow the protocol used for FISH.

1. Prior to PRINS, wash slides three times in 2× SSC for 5 min at RT.
2. Treat slides with RNase (50 µg/mL in 2× SSC) for 40 min at 37°C. Subsequently a fill-in reaction is carried out to reduce background signals

caused by nicks within the chromosomal DNA which may induce polymerase activity at sites of free 3′ OH ends.

3. To reduce background signals, wash slides first in 2× SSC and equilibrate
   in *Taq*-polymerase buffer (10 mM Tris-HCl, pH 8.4, 50 mM KCl, 1.5 mM
   $MgCl_2$) for 5 min.

4. Reaction mixture (20 μL)
   a. 1× PCR buffer (Boehringer)
   b. mM $MgCl_2$
   c. 100 μM of each dATP, dCTP, dGTP
   d. 100 μM of 2′,3′-dideoxy (dd)TTP
   e. 2U of *Taq*-DNA polymerase (Boehringer)

5. Drop reaction mixture on slide, cover with a cover slip and seal with
   Fixogum rubber cement.

6. Heat slides at 93°C for 90 s followed by 72°C for 20 min in a wet chamber
   (Zytotherm, Schutron). Before labeling, wash chromosomes and equilibrate
   again.

7. Labeling mixture (25 μL)
   a. 1× PCR buffer
   b. mM $MgCl_2$
   c. 100 μM of each dATP, dCTP, dGTP, 75 μM dTTP
   d. 25 μM of digoxygenin-11-dUTP or fluorescein-12-dUTP (Boehringer)
   e. 4 μM of each of the corresponding oligonucleotide primers and 2.5 U
      of *Taq*-DNA polymerase

8. Seal the cover slip, denature chromosomes for 2–3 min at 93°C. Anneal
   primers at 55°C–60°C for 10 min and extend at 72°C for 40 min.

9. Wash slides twice in 4× SSC, 0.1% Tween 20 for 5 min at 42°C in order to
   stop the reaction.

10. Counterstain preparation labeled with FITC-dUTP with propidium
    iodide/4′, 6-diamidino-2-phenylindole dihydrochloride (PI/DAPI, 1 μg/mL
    each in antifade) and examine immediately.

11. Detect DIG-dUTP with antidig-FITC-Fab fragments from sheep (2 μg/mL,
    Boehringer).

12. Incubate samples in blocking solution (4× SSC, 0.1% Tween 20, 3% BSA) at
    37°C for 30 min.

13. Subsequently, apply Fab fragments in detection buffer (4× SSC, 0.1%
    Tween 20, 1% BSA) for 50 min at 37°C.

14. Remove unspecifically bound conjugates by washing slides in wash buffer
    three times for 5 min at 42°C.

15. After counterstaining, described above, examine slides using Zeiss epifluo-
    rescence microscope.

4.11.3   FISH ON EXTENDED DNA FIBER: FIBER-FISH

Fiber-FISH technique is a very powerful cytological tool to analyze large repetitive
regions and increases the resolution of FISH analyses down to a few kb in the higher
eukaryotic genomes (http://jianglab.horticulture.wisc.edu/Resource_files/Fiber%20

FISH%20Protocols.pdf). It can be used to gage the distances between adjacent clones up to ~500 kb and to measure repetitive loci up to ~1.7 mb. Combined with metaphase and interphase nuclei analysis this tool allows molecular cytogeneticists to map loci to specific chromosomes and determine the distance between loci from a few kb up to several mb.

4.11.3.1 Protocol-I: Fiber-FISH on Extended Nuclear DNA Fibers

4.11.3.1.1 Isolation of Plant Nuclei

Isolation of plant nuclei is the same as for the preparation of high molecular weight DNA. The nuclei are suspend in 50% glycerol to store at −20°C. Almost all of the published protocols call for a 20 mm or 30 μm filtrations, but this can be omitted it with most of plant materials in order to obtain the maximum number of nuclei. If there is too much debris in the suspension or on slides the last (20 μm) filtration can be added.

1. Freeze 2–5 g fresh leaf material in liquid nitrogen and grind to a fine powder with a precooled (−20°C) mortar and pestle.
2. Transfer powder to a 50 mL centrifuge tube, add 20 mL chilled nuclei isolation buffer [NIB] and mix gently (make sure to break-up clumps) on ice for 5 min (in an ice bucket on a shaker). [NIB: 10 mM Tris-HCl pH 9.5, 10 mM EDTA, 100 mM KCl, 0.5 M sucrose, 4.0≈mM spermidine, 1.0 mM spermine, 0.1% mercaptoethanol. Prepare a large stock and store it in a refrigerator (4°C). Mercaptoethanol should not be included in the stock. It can be added just before use.]
3. Filter through nylon mesh: 220 μm, 148 μm, 48 μm, and 22 μm sequentially, into cold (on ice) 50 mL centrifuge tubes using chilled funnel. (Nylon filters obtained from: Tetko Inc, P.O. Box 346, Lancaster, NY, 14086, tele 914-941-7767)
4 Add 1 mL NIB containing 10% (v/v) Triton X-100 (premixed) and gently mix the filtrate. The final concentration of Triton X-100 should be 0.5%. It removes any chloroplast contamination. Centrifuge at 2000 × g for 10 min at 4°C, Decant the supernatant. If pellet is very small, skip the further cleaning steps and move directly to step vii. Otherwise, resuspend the large pellet in 20 mL NIB (with mercaptoethanol added).
5. Filter through nylon mesh: 48 μm and 30 μm sequentially (optional), into cold (on ice) 50 mL centrifuge tubes.

   Note: A 22 μm filter step will lose a lot of nuclei. However, this makes the nuclei cleaner.

6. Add 1 mL NIB containing 10% (v/v) Triton X-100 (pre-mixed) and gently mix the filtrate, as in step iv. Centrifuge at 2000 × g for 10 min at 4°C.
7. Decant the supernatant and resuspend the pellet in 200 μL to 5 mL of 1:1 NIB: 100% glycerol (neither mercaptoethanol nor Triton X-100 added) depending on the amount of nuclei harvested (concentration ~5 × 10^6 nuclei/mL, can be checked by staining with DAPI and examining under a microscope). Store at −20°C.

4.11.3.1.2   Extension of DNA Fibers

Extending the fibers is a critical step in fiber-FISH. There are several methods of extending the fibers. The dragging method with a coverslip seems to give the most uniform results. When dragging it is imperative that it be done slowly and smoothly. Poly-L-lysine slides obtained from Sigma can be used. These slides are treated so as to promote the adhesion of one or both ends of the DNA molecule. Silinated slides can also be used but seem to generate too much background signal. The calibration of the method should be checked occasionally by using BACs or cosmids of a known length as probes.

1. Identify the nuclei portion in the nuclei stock. The nuclei tend to settle near the bottom of the tube and the settling process can take a day or longer. The pellet may appear white and the nuclei often times sit right above this bottom film. The color of the nuclei is variable across species and samples, but normally clean nuclei has a gray/white coloration. Any layers above the nuclei tend to contain debris. Some people like to mix the nuclei stock prior to slide preparation by gently inverting the Eppendorf tube several times. It is not desirable as it mixes the debris with the nuclei.
2. With a cut P20 pipette tip, pipette 1–10 µL nuclei suspension (1–5 µL/slide depending on the suspension concentration) into ~100 µL NIB (minus mercaptoethanol and Triton) in an Eppendorf tube to dilute the glycerol. Gently mix the nuclei with the buffer and centrifuge at 3000–3600 rpm for 5 min. Remove carefully the supernatant with a pipette leaving only the nuclei pellet.
3. Resuspend the nuclei in PBS (the final volume is 2 µL per slide) (PBS: 10 mM sodium phosphate, pH 7.0, 140 mM NaCl)
4. Pipette 2 µL suspension across one end of a clean poly-L-lysine slide (Sigma, Poly-Prep, Cat # P0425) and air dry for 5–10 min. The nuclei should dry to the point where it appears sticky: neither wet nor dry.
5. Pipette 8 µL STE lysis buffer on top of the nuclei and incubate at RT for 4 min (STE: 0.5% SDS, 5 mM EDTA, 100 mM Tris, pH 7.0)
6. Slowly drag the solution down the slide with the edge of a clean coverslip held just above the surface of the slide, do not touch the coverslip to the slide surface as this will drag the nuclei completely off the slide. Air dry for 10 min.
7. Fix in fresh 3:1 100% ethanol:glacial acetic acid for 2 min.
8. Bake at 60°C for 30 min.
9. Slides can be used immediately but it can be stored in a box for several weeks, but it is suggested to use them immediately.

4.11.3.1.3   Probe Application

1. Apply 10 µL probe to the slide, then cover with a 22 mm × 22 mm cover slip and seal with rubber cement.
2. After the cement is dried, place the slide in an 80°C oven for 3 min in direct contact with a heated surface, then for 2 min in a wet chamber prewarmed in the 80°C oven.
3. Transfer wet chamber, with the slides, immediately to 37°C overnight. It is recommended to incubate at 37°C for longer periods, up to three or four days, especially for difficult probes.

4.11.3.1.4   Probe Detection

Three-layer detection gives a much stronger signal than does the single layer of antibodies. All antibody layers are composed of the antibodies diluted in the appropriate buffers at the concentration specified below:

Apply 100 µL antibody to each slide and a 22 mm × 40 mm coverslip is gently placed upon the antibody solution to promote even spreading. All antibody layers are incubated in a 37°C wet chamber for a minimum of 30 min. The first layer is often incubated for up to 45–60 min.

Notes: The blocking step using 4M buffer seems to help reduce some of the background noise. Dry bovine milk from Sigma works the best in the 4M buffer, other substitutes (i.e., Carnation dry milk) tend to reduce the amount of signal. The 4M and TNB buffers can be prepared at 5× and stored at −20°C. The wash solutions, 4T and TNT, can be prepared at 20× and 10×, respectively, and stored at RT.

i. *One-color detection protocol*

	Time, min (total)
• Wash in 2× SSC	5
Wash in 2× SSC 42°C	10
Wash in 2× SSC	5
• Wash in 1× 4T	5
Incubate at 37°C in 4M	30
Wash in 1× 4T	2
• Incubate FITC–avidin (1 µl antibody stock/100 µL TNB buffer)	30
Wash 3 times in 1× TNT	5 (15)
• Incubate biotin anti-avidin (0.5 µL/100 µL TNB buffer)	30
Wash 3 times in 1× TNT	5 (15)
• Incubate FITC–avidin (1 µL/100 µL TNB buffer)	30
• Wash 3 times in 1× TNT	5 (15)
Wash 2 times in 1× PBS	5 (10)

• Add 10 µL Prolong (Molecular Probes) or Vectashield (Vector Labs), cover with a 22 mm × 30 mm cover slip and squash

Notes:

• All 30 min incubation periods are at 37°C.
• Antibodies:
  – FITC–avidin, 1 µL per 100 µL buffer
  – Biotin antiavidin, 0.5 µL per 100 µL buffer
  – Mouse anit-dig, 1 µL per 100 µL buffer
  – DIG antimouse, 1 µL per 100 µL buffer
  – Rodamine antidig, 1–2 µL per 100 µL buffer

- Solutions:
  - *4M:* 3%–5% nonfat dry milk (Sigma, Cat # M7409) in 4× SSC
  - *4T:* 4× SSC, 0.05% Tween 20
  - *TNB:* 0.1 M Tris-HCl, pH 7.5, 0.15 M NaCl, 0.5% blocking reagent (Boehringer Mannheim)
  - *TNT:* 0.1 M Tris-HCl, 0.15 M NaCl, 0.05% Tween 20, pH 7.5
  - *PBS:* 0.13 M NaCl, 0.007 M Na_2HPO_4, 0.003 M NaH_2PO_4

ii. *Two-color detection protocol*

	Time, min (total)
• Wash in 2× SSC	5
Wash in 2× SSC 42°C	10
Wash in 2× SSC	5
• Wash in 1× 4T	5
Incubate at 37°C in 4M	30
Wash in × 4T	2
• Incubate FITC anti-dig (1 µL antibody stock /100 µL TNB buffer) + Texas Red streptavidin (1 µL/100 µL), 37°C for 30 min	
Wash 3 times in 1× TNT	5 (15)
• Incubate FITC anti-sheep (1 µL/100 µL TNB buffer) + biotinylated anti-avidin (1 µL/100 µL), 37_C for 30 min	
Wash 3 times in 1× TNT	5 (15)
• Incubate Texas Red streptavidin (1 µL/100 µL TNB buffer), 37°C for 30 min.	
Wash 3 times in 1× TNT	5 (15)
Wash 2 times in 1× PBS	5 (10)
• Add 10 µL Vectashield (Vector Labs), cover with a 22 mm x 30 mm cover slip and squash	

Notes: Preparation of antibody stocks:

- FITC antisheep comes ready to use and is stable at 4°C
- FITC antidig needs 1 mL water resuspension.

iii. *Poly-L-lysine slide preparation*
   A. Boil slides in 5 M HCl for 2–3 h
   B. Rinse thoroughly with dH_2O then air dry
   C. Incubate overnight in filtered 10^{-6} g/mL poly-*D*-lysine (MW = 350,000, Sigma)
   D. Rinse thoroughly

iv. *Slide silanation*
  A. 30 min 1:1 HCl:methanol
  B. Overnight 18 M sulfuric acid
  C. 8–10 washes in ddH_2O
  D. 10 min boiling ddH_2O
  E. 1 h 10% 3-aminopropyltriethoxy silane in 95% ethanol
  F. Rinse several times in ddH_2O
  G. Wash in 100% ethanol
  H. 80°C–100°C overnight before use

Note: The Poly-Prep slide from Sigma (Cat # P 0425) is ready to use is available and can be used instead of the above procedure.

4.11.3.2   Protocol 2: Fiber-FISH Using BAC and Circular Molecules as Targets

1. *Prior to preparing slides*
   a. Label appropriate probes (biotin and dig).
   b. Miniprep BAC DNA (Solution I, II, III method followed by IPA precipitation; use 20 µL water for resuspension).
   c. Silanize 22 mm × 22 mm cover slips by dipping in Sigmacote for 10 min, then air dry.
2. *Slide preparation*
   a. Prepare wet chamber at 37°C, turn slide warmer up to 60°C.
   b. Dilute BAC DNA (w/cut P20 pipette tips) to appropriate level (dilute 1 µL BAC into 9 µL water). Add all 10 µL of diluted BAC to Poly-Prep slide (Sigma #P0425).
   c. Add 15 µL of **FISH lysis buffer*** to BAC drop. Allow drop to spread. Let this sit at RT for ~5 min. Add water to the slide if it dries.
   d. Gently place ("drop") a silanized cover slip directly over the liquid (use tweezers to avoid air bubbles).
   e. Transfer slides to slide warmer. Allow slides to "bake" for 15 min. At this point, one should see the liquid begin to recede.
   f. Place slides in 3:1 (ethanol: glacial acetic acid), wait 1 min; gently shake slide to promote removal of the cover slip. Once cover slip falls off, transfer slides to new container of 3:1 and incubate for 1 min 30 s. Transfer slides back to slide warmer for an additional 15 min.
   g. Add probe, denature, and detect as in nuclear fiber-FISH.
   *FISH lysis buffer
   – 2% Sarkosyl
   – 0.25% SDS
   – 50 mM Tris (pH 7.4)
   – 50 mM EDTA (pH 8.0)

4.11.3.3   Protocol III: Staining Fibers (Yo-Yo Staining)

   a. Prepare slides for fiber analysis (as in BAC or nuclear fiber-FISH protocols).

b. Dilute Yo-Yo (molecular probes) in PBS following manufacturer's direction.

c. Add 100 μL of the Yo-Yo dilution to the slide and add a cover slip. Store in a dark place at RT for 10–20 min.

d. Wash three times in PBS (5 min each).

e. Short dry and add antifade (Vectashield).

f. View slides.

4.11.3.4   Source of Chemicals

Appendix 2-II

REFERENCES

Brammer, S. P., S. Vasconcelos, L. B. Poersch, A. R. Oliveira, and A. C. Brasileiro-Vidal. 2013. Genomic in situ hybridization in Triticeae: A methodological approach. In: *Agricultural and Biological Sciences-Plant Breeding from Laboratories to Fields.* S. B. Andersen (ed.). ISBN 978-953-51-1090-3.

Endo, T. R. 2011. Cytological dissection of the Triticeae chromosomes by the gametocidal system. In: *Plant Chromosome Engineering: Methods and Protocols, Methods in Molecular Biology.* J. A. Birchler (Ed.). Springer Science + Business media, LLC. pp. 247–257.

Fukui, K., N. Ohmido, and G. S. Khush. 1994. Variability in rDNA loci in genus *Oryza* detected through fluorescence in situ hybridisation. *Theor. Appl.* Genet. 87: 893–899.

Fukui, K., and S. Nakayama. 1996. Plant Chromosomes: Laboratory Methods. (Eds.) K. Fukui and S. Nakayama. CRC Press, Inc., Boca Raton, FL.

Gall, J. G., and M. L. Pardue. 1969. Formation and detection of RNA-DNA hybrid molecules in cytological preparations. *Proc. Natl. Acad. Sci. U. S. A.* 63: 378–383.

Gosden, J. R. 1997. *PRINS and In Situ PCR Protocols.* Humana Press, Totowa, NJ.

Kato, A. 2011. High-density fluorescence in situ hybridization signal detection on barley (*Hordeum vulgare* L.) chromosomes with improved probe screening and reprobing procedures. *Genome.* 54: 151–159.

Koch, J. E., S. Kølvraa, K. B. Petersen, N. Gregersen, and I. Bolunel. 1989. Oligonucleotide-priming methods for the chromosome-specific labelling if alpha satellite DNA in situ. *Chromosoma* 98: 259–265.

Komuro, S., R. Endo, K. Shikata, and A. Kato. 2013. Genomic and chromosomal distribution patterns of various repeated DNA sequences in wheat revealed by a fluorescence in situ hybridization procedure. *Genome.* 56:131–137.

Koo, D. H., S. K. Sehgal, B. Friebe, and B. S. Gill. 2015. Structure and stability of telocentric chromosomes in wheat. *Plos ONE.* 1–16. DOI:10.1371/journal.pone.0137747.

Menke, M., J. Fuchs, and I. Schubert. 1998. A comparison of sequence resolution on plant chromosomes: PRINS versus FISH. *Theor. Appl. Genet.* 97: 1314–1320.

Mukai, Y. 1996. Multicolor fluorescence *in situ* hybridization: a new tool for genome analysis. In *Methods of Genome Analysis in Plants.* P. P. Jauhar (Ed.). CRC Press, Boca Raton, FL. pp. 181–192.

Pellestor, F., I. Quennesson, L. Coignet, A. Girardet, B. Andréo, and J. P. Charlieu. 1996. Direct detection of disomy in human sperm by the PRINS technique. *Human Genet.* 97: 21–25.

Roche, P. J. 2006. Preparation of template DNA and labeling techniques. 2006. In: *Methods in Molecular Biology,* vol. 326: *In Situ* Hybridization Protocols: 3rd Ed. I. A. Darby and T. D. Hewitson (Eds.). Humana Press Inc., Totowa, NJ, pp. 9–16.

Shishido, R., S. Apisitwanich, N. Ohmido, Y. Okinaka, K. Mori, and K. Fukui. 1998. Detection of specific chromosome reduction in rice somatic hybrids with the A, B, and C genomes by multi-color genomic *in situ* hybridization. *Theor. Appl. Genet.* 97: 1013–1018.

Singh, R. J. 2017. *Plant Cytogenetics.* 3rd Ed. CRC Press, Boca Raton, FL.

Singh, R. J., H. H. Kim, and T. Hymowitz. 2001. Distribution of rDNA loci in the genus *Glycine* willd. *Theor. Appl. Genet.* 103: 212–218.

Wang, K., W. Zhang, Y. Jiang, and T. Zhang. 2013. Systematic application of DNA fiber-fish technique in cotton. *PLoS ONE* 8(9): e75674. DOI:10.1371/ journal.pone.0075674.

5 Flow Analysis and Sorting of Plant Chromosomes

5.1 INTRODUCTION

Methods of chromosome analysis and sorting by flow cytometry (flow cytogenetics) are increasingly used in plant genomics. Protocols for flow cytometry are provided by J. Doležel (personal communication) which was published by Doležel et al. (2014) and Vrána et al. (2016). Flow cytometry can be used for rapid classification of isolated chromosomes according to their relative DNA content. The resulting distributions, generally referred to as flow karyotypes, enable detection of numerical and structural chromosome aberrations. Flow-sorted chromosomes are an invaluable source of DNA from small and defined parts of genomes; their use offers a lossless reduction of genome complexity to simplify the analysis and mapping of complex plant genomes. For instance, flow-sorted chromosomes may be used for physical gene mapping, isolation of molecular markers from particular chromosomes, construction of chromosome-specific DNA libraries and sequencing. List of all plant species, where flow cytometric analysis and sorting experiments have been published (Table 5.1). A general outline of the procedure for flow cytometric analysis and sorting of plant chromosomes consists of the following steps: (1) induction of cell cycle synchrony and accumulation of dividing cells in metaphase, (2) preparation of suspensions of intact chromosomes, (3) flow analysis and sorting, and (4) processing of flow-sorted chromosomes.

5.2 BASIC PROTOCOL 1

5.2.1 ACCUMULATION OF ROOT-TIP CELLS IN MITOTIC METAPHASE

- Induces a high degree of cell cycle synchrony in meristem root-tip cells and accumulates dividing cell in metaphase.
- This procedure consists of successive treatments of actively growing roots with hydroxyurea (a DNA synthesis inhibitor) and anti-microtubular drugs amiprophos-methyl and oryzalin.
- Optimum concentrations and treatment times are determined experimentally for each species, vary from species to species (Table 5.2).
- Adjust the temperature of all solutions to $25 \pm 0.5°C$ prior to use.
- Perform all incubation in the dark in a biological incubator at $25 \pm 0.5°C$.
- Aerate all solutions.
- Keep aeration stones and tubing clean to avoid extensive contamination by bacteria and fungi.

TABLE 5.1
List of Plant Species with Published Protocols for Flow Cytometric Experiments with Metaphase Chromosomes

Species	Common Name	References
Aegilops spp. (*biuncialis,comosa, geniculata, umbellulata*)	Goatgrass	Molnár et al. (2011)
Aegilops spp. (*speltoides, tauschii*)		Molnár et al. (2014); Akpinar et al. 2015b)
Aegilops spp. (*cylindrica, markgrafii, triuncialis*)		Molnár et al. (2015)
Avena sativa	Oat	Li et al. (2001)
Cicer arietinum	Chickpea	Vláčilová et al. (2002) Zatloukalová et al. 2011)
Cunninghamia lanceolata	Chinese fir	Shen et al. (2015)
Dasypyrum villosum	Mosquito grass	Grosso et al. (2012); Giorgi et al. (2013)
Festuca pratensis	Meadow fescue	Kopecký et al. (2013)
Haplopappus gracilis	Slender goldenweed	de Laat and Blaas (1984); de Laat and Schel (1986)
Hordeum vulgare	Barley	Lysák et al. (1999); Lee et al. (2000); Suchánková et al. (2006)
Lycopersicon esculentum	Tomato	Arumuganathan et al. (1991)
Lycopersicon pennellii	Tomato	Arumuganathan et al. (1991, 1994)
Nicotiana plumbaginifolia	Tobacco	Conia et al. (1989)
Oryza sativa	Rice	Lee and Arumuganathan (1999)
Petunia hybrida	Petunia	Conia et al. (1987)
Picea abies	Norway spruce	Überall et al. (2004)
Pisum sativum	Pea	Gualberti et al. (1996); Neumann et al. (1998, 2002)
Silene latifolia (syn. *Melandrium album*)	White campion	Veuskens et al. (1995); Kejnovsk´y et al. (2001); Kr´alov´a et al. (2014)
Triticum aestivum	Bread wheat	Wang et al. (1992); Lee et al. (1997); Schwarzacher et al. (1997); Gill et al. (1999); Vrána et al. (2000); Kubaláková et al. (2002)
Triticum dicoccoides	Durum wheat	Kubaláková et al. (2005); Giorgi et al. (2013)
Triticum urartu	Wild einkorn	Molnár et al. (2014)
Vicia faba	Field bean	Lucretti et al. (1993); Doležel and Lucretti (1995); Lucretti and Doležel (1997)
Vicia sativa	Common vetch	Kovářová et al. (2007)
Zea mays	Corn	Lee et al. (1996, 2002); Li et al. (2001, 2004)

TABLE 5.2

Protocols for Preparation of Suspensions of Plant Metaphase Chromosomes

Species	Hydroxyurea Treatment Concentration (mM)	Duration (h)	Recovery Time Duration (h)	Metaphase Accumulation Agent and Concentration	Duration (h)	Formaldehyde Fixation Concentration (%)	Duration (min)	Tissue Homogenization RPM	Duration (s)
Aegilops species	1.25	18	5	APM (2.5µM)	2+ice overnight	2	20	20,000	13
Agropyron cristatum	1.25	18	4.5	APM (2.5µM)	2+ice overnight	2	15	18,000	13
Arachis hypogea	3.00	18	5	APM (5µM)	2	3	20	15,000	18
Asparagus officinalis	3.00	18	6	APM (5µM)	2	2	20	15,000	13
Avena sativa	2.00	18	4.5	Oryzalin (10µM)	2+ice overnight	2	25	25,000	13
Avena strigosa	2.00	18	5	Oryzalin (5µM)	2	2	20	20,000	13
Cicer arietinum	1.25	18	4	Oryzalin (5µM)	2+ice overnight	2	30	10,000	18
Crocus sativus	0.5	18	5	APM (5µM)	3+ice overnight	2	20	Chopping	
Dasypyrum villosum	1.25	18	5	APM (2.5 µM)	2+ice overnight	2	20	20,000	13
Elymus elongatum	1.25	18	5	N₂O (506.625kPa)	2+ice overnight	2	20	20,000	13
Festuca pratensis	1.0	18	5	APM (2.5 µM)	2	2	20	20,000	15
Gossypium hirsutum	4.5	18	3.5	Oryzalin (5µM)	2+ ice overnight	2	25	13,000	18
Hordeum vulgare	2.0	18	6.5	APM (2.5µM)	2+ice overnight	2	20	15,000	13
H. vulgare cv. Barke	1.5	18	5.5	APM (2.5µM)	2+ice overnight	2	20	15,000	13
Lens culinaris	1.25	18	4.5	APM (10 µM)	2+ice overnight	2	20	13,000	18
Lolium perenne	1.5	18	4.5	APM (5µM)	2	2	20	20,000	15
Lupinus angustifolius	3.0	18	4	Oryzalin (2.5µM)	2+ice overnight	2	25	18,000	18
Medicago sativa	4.5	18	6	APM (2.5µM)	3	2	20	9,500	10
Pisum sativum	1.25	18	4.5	APM (10 µM)	2+ice overnight	2	20	13,000	18
Rumex acetosa	2.0	18	5	Oryzalin (10µM)	2+ice overnight	2	20	18,000	13

(Continued)

TABLE 5.2 (Continued)
Protocols for Preparation of Suspensions of Plant Metaphase Chromosomes

Species	Hydroxyurea Treatment Concentration (mM)	Hydroxyurea Treatment Duration (h)	Recovery Time Duration (h)	Metaphase Accumulation Agent and Concentration	Metaphase Accumulation Duration (h)	Formaldehyde Fixation Concentration (%)	Formaldehyde Fixation Duration (min)	Tissue Homogenization RPM	Tissue Homogenization Duration (s)
Saccarum officinarum	2.0	18	4.5	APM (2.5μM)	2	2	20	18,000	18
Secale cereale	2.5	18	6.5	APM (10μM)	2+ice overnight	2	25	15,000	13
Silene species[a]	2.0	18	5	Oryzalin (2.5μM)	2	2	15	18,000	13
Triticum aestivum	2.0	18	5.5	APM (2.5μM)	2+ice overnight	2	20	20,000	13
Triticum species[b]	1.25	18	5	APM (2.5μM)	2+ice overnight	2	20	20,000	13
Vicia faba	1.25	18.5	4.5	APM (2.5μM)	2	4	30	Chopping or 15,000	18
Vicia sativa	2.5	18.5	3.5	Oryzalin (2.5μM)	2	2	25	13,000	18
Vigna unguiculata	1.25	18	4	Oryzalin (2.5μM)	2+ice overnight	2	20	13,000	18
Zea mays	3.0	18	3.5	N₂O 506.625kPa	3	3	25	Chopping	

[a] S. latifolia, S. dioica;
[b] T. dicoccoides, T. durum, T. militinae, T.monococcum, T. urartu

5.2.2 MATERIALS

- Deionized water (dH_2O)
- Seeds
- Hydroxyurea treatment solution
- 1× or 0.1× Hoagland's nutrient solution
- Amiprophos-methyl treatment solution
- Oryzalin
- 18 cm diameter glass Petri dish with a lid
- Filter paper cut to 18 cm diameter
- Biological incubator (heating/cooling) with internal temperature adjusted to $25 \pm 0.5°C$
- 750 mL plastic tray (e.g., 14 cm long × 8 cm wide × 10 cm high) including an open-mesh basket to hold germinated seeds
- Aquarium bubbler with tubing and aeration stones
- Iced water bath

5.2.3 GERMINATION OF SEEDS

1. Place several layers of paper towels into an 18 cm glass Petri dish and top them with a single sheet of filter paper.
2. Moisten the paper layers with dH_2O.
3. Spread seeds on the filter paper surface.
   Approximately 20–50 seedlings (according to the number of roots per seedling and size of root meristem) are needed to prepare one sample (1 mL of chromosome suspension prepared in basic protocol 2).
4. Cover the Petri dish with a glass lid and germinate the seeds at $25 \pm 0.5°C$ in a biological incubator in the dark.
   Optimal root length is 2–3 cm and should be achieved in 2–3 days.
5. Select seedlings with primary roots of similar length.
6. Thread seedlings roots through the holes of an open-mesh basket placed in a 750 mL plastic tray filled with dH_2O (Figure 5.1).

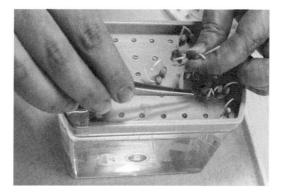

FIGURE 5.1 Germination and collection of roots. (From Doležel et al. *Biotechnology Advances* 32, 122–136, 2017.)

5.2.4   Accumulation of Root-Tip Cells in Metaphase

7. Transfer the basket with seedlings to a second plastic tray containing hydroxyurea treatment solution. Incubation times and hydroxyurea concentrations of each species are indicated in Table 5.1.
8. Transfer the basket with seedlings to tray filled with either amiprophosmethyl or oryzalin treatment solution. For concentration and incubation times see Table 5.2.
9. Incubate in hydroxyurea-free Hoagland's nutrient solution for recovery time specified in Table 5.1.
10. Transfer the basket with seedlings to a tray filled with either amiprophosmethyl or oryzalin treatment solution. For concentration and incubation times see Table 5.2.

    In some plant species of overnight treatment in ice water (step 11 and 12) helps to reduce frequency of chromosome clumps.

11. Transfer the basket with seedlings to a plastic tray filled with an ice-water bath (1°C–2°C).
12. Place the container in a refrigerator and leave overnight.

    To determine the degree of synchrony, see support protocol 1. Seedlings should not be stored before chromosome suspensions are prepared.

5.3   ALTERNATE PROTOCOL 1

5.3.1   Accumulation of Root-Tip Cells in Mitotic Metaphase in Large-Seeded Legumes

Modification of basic protocol 1 for large-seeded legumes such as field bean, chickpea, and garden pea.

5.3.2   Additional Materials

- Also see basic protocol 1
- Insert substrate for seed germination (e.g., perlite)
- Four liter plastic tray (e.g., 25 cm long × 15 cm wide × 11 cm high)
- Aluminum foil

1. Soak seeds for 24 h in dH₂O with aeration. Approximately 30 seedlings are needed to prepare one sample (1 mL of chromosome suspension prepared in basic protocol 2).
2. Wet an inert substrate (e.g., perlite) with 1× Hoagland's nutrient solution and put it into a 4 L plastic tray.
3. Wash seeds in 1000 mL dH₂O, spread them over the surface of the wet substrate, and cover them with a 1 cm layer of wet substrate.
4. Cover the tray with aluminum foil and germinate the seeds at 25 ± 0.5°C in a biological incubator in the dark.

   Optimal root length is ~4 cm and should be achieved in 2–3 days.

5. Remove seedlings from the substrate and wash them vigorously in 1000 mL dH$_2$O.
6. Continue with metaphase accumulation procedure as described in basic protocol 1, steps 7–10. In case of chickpea and garden pea, follow the steps 11–12 as well.

5.4 ALTERNATE PROTOCOL 2

5.4.1 ACCUMULATION OF ROOT-TIP CELLS AT THE MITOTIC METAPHASE BY NITROUS OXIDE TREATMENT

This protocol is suitable for plant species, where the standard method of metaphase accumulation using mitotic poisons causes chromosome clumping, which hampers successful chromosome isolation.

5.4.2 ADDITIONAL MATERIALS

- Also see basic protocol 1
- Pressure chamber and nitrous oxide (N$_2$O) cylinder

1. Follow the steps 1–9 of the basic protocol 1.
2. Place the basket with seedlings into the pressure chamber.
3. Seal the chamber by tightening nuts holding the pressure chamber doors. Set the pressure of N$_2$O to 506.625 k Pa and incubate the seedlings for 2 h (meadow fescue) or 3 h (maize).

5.5 ALTERNATE PROTOCOL 3

5.5.1 ACCUMULATION OF ROOT-TIP CELLS AT THE MITOTIC METAPHASE IN PLANT SPECIES WHERE SEEDS ARE UNAVAILABLE

This protocol is suitable for plant species where seeds are not available. Basic protocol 1 has been adapted to bulbs of crocus or fresh stem cuttings of sugar cane.

5.5.2 ADDITIONAL MATERIALS (ALSO SEE BASIC PROTOCOL 1)

- Wire netting covers with holes suitable for holding crocus bulbs

5.5.3 GROW THE ROOTS

1a. *Crocus:*

- Put the bulbs into wet substrate (one third perlite, one third sand, one third garden soil) and keep them at 20°C until the roots achieve 2–3 cm in length.
- Remove the bulbs or stem cuttings from the substrate and thoroughly wash the roots in deionized water.

1b. *Sugar cane:*

- Put the fresh stem cuttings (2–3 cm in diameter) into wet substrate (perlite) at 20°C until the roots achieve 2–3 cm in length.
- Remove the bulbs or stem cuttings from the substrate and thoroughly wash the roots in dH_2O.

5.5.4 PREPARE FOR TREATMENTS

2a. *Crocus:*

- Position the bulbs on the top of the wire netting cover of the plastic tray by threading the roots through the holes.

2b. *Sugar cane:*

- Put the cuttings into the plastic tray in such a way that the roots head toward the bottom.
- Continue with the synchronization/metaphase accumulation procedure according to steps 7–12 of the basic protocol 1. Fill the plastic trays with appropriate treatment solutions until the roots are fully immersed.
  In case of sugar cane, increase the temperature of all synchronization treatments to 30°C.

5.6 SUPPORT PROTOCOL 1

5.6.1 ANALYSIS OF THE DEGREE OF METAPHASE SYNCHRONY

This support protocol is used to estimate the frequency of metaphase cells following the procedures for metaphase accumulation (basic protocol 1 and alternate protocols 1, 2, and 3).

5.6.2 MATERIALS

- Root tips synchronized in metaphase (see basic protocol 1, alternate protocols 1, 2, or 3)
- Deionized water (dH_2O)
- 3:1 (v/v) ethanol/glacial acetic acid, freshly prepared
- 70% and 96% (v/v) ethanol
- 5N HCl
- Schiff's reagent (see recipe)
- 45% (v/v) acetic acid
- Fructose syrup (see recipe)
- DePeX mounting medium (Serva)
- 4°C incubator
- Microscope slides and coverslips

- 18 mm × 18 mm coverslips
- Coplin staining
- Microscope

5.6.3 FIX AND STAIN CELLS

1. Harvest 1 cm root tips in dH$_2$O
   - About 10 root tips should be used to allow for accidental losses during the procedure and for low-quality preparations.
   - At least five good squashes should be used for microscopic evaluation.
2. Fix in 3:1 ethanol/glacial acetic acid overnight at 4°C.
   CAUTION: *Glacial acetic acid is volatile. Concentrated acids must be handled with great care. Wear gloves and safety glasses and work in a chemical fume hood.*
3. Remove fixative with several washes in 50 mL of 70% ethanol.
   Fixed tips may be stored in 70% ethanol up to 1 year at 4°C.
4. Wash tips in several changes of dH$_2$O.
5. Hydrolyze tips in 5N HCl for 25 min at room temperature
   CAUTION: *Concentrated HCl is volatile. Concentrated acids should be handled with great care. Wear gloves and safety glasses and work in a chemical fume hood.*
6. Wash in dH$_2$O and incubate in Schiff's reagent for 1 h at room temperature.

5.6.4 PREPARE AND ANALYZE SQUASHES

7. Wash tips in dH$_2$O and macerate (soften) them for about 1 min in 45% acetic acid at room temperature.
   - *To make slides for immediate use*
8a. Cut off the darkly stained meristem tip and squash it in a drop of fructose syrup between a microscope slide and an 18 mm × 18 mm coverslip. Repeat to prepare at least five different slides.
8b. Squash the darkly stained meristem tip in a drop of 45% acetic acid and immediately place the slide on a block of dry ice. Allow the slide to freeze. Repeat to prepare at least five different slides.
9a. To each of five slides, analyze ≥1000 cells and determine the proportion of cells in metaphase.
   Squash preparations in fructose syrup can be maintained for a few days in a refrigerator, but are not permanent.
   - *To make permanent slides*
9b. Peel off coverslips. Dehydrate tips in two changes of 96% ethanol in Coplin jars and leave to air dry overnight.
10. Mount in a drop of DePeX.
11. On each of five slides, analyze ≥1000 cells and determine the proportion of cells in metaphase.

5.7   BASIC PROTOCOL 2

5.7.1   PREPARATION OF SUSPENSIONS OF INTACT PLANT CHROMOSOMES

- This protocol is based on the method developed by Doležel et al. (1992).
- After a mild fixation with formaldehyde, chromosomes are mechanically released from synchronized root tips into a polyamine buffer (LB01), which stabilizes their structure.
- Chromosome suspensions prepared according to this procedure are suitable for flow cytometric analysis and sorting (see basic protocols 3 and 4).

5.7.2   MATERIALS

- Synchronize root tips in metaphase (see basic protocol 1 and alternate protocols 1, 2, or 3)
- Deionized water (dH$_2$O)
- Formaldehyde fixative (see recipe)
- Tris buffer (see recipe)
- LB01 lysis buffer (see recipe)
- LB01-P lysis buffer (see recipe)
- Ice
- Isolation buffer, 1.5× (1.5× IB, see recipe)
- mg/mL DAPI stock solution (see recipe)
- Cooled water bath (5°C)
- 5 mL polystyrene tubes (e.g., Falcon 352008; Corning)
- Mechanical homogenizer (e.g., Polytron PT1300D with a PT-DA 1205/5 probe; Kinematica AG)
- 50 μm (pore size) nylon mesh in 4 cm × 4 cm squares
- 0.5 mL tubes for polymerase chain reaction (PCR)
- Microscope slides
- Fluorescence microscope with 10× to 20× objective, UV light source, and DAPI filter set

5.7.3   PREPARE CHROMOSOME SUSPENSION

1. Harvest 1 cm root tips and transfer into dH$_2$O.
   Root tips must be harvested immediately after metaphase arrest treatment (see basic protocol 1, step 10) or ice water (see basic protocol 1, step 12).
2. Immediately transfer root tips to 25 mL formaldehyde fixative and fix at 5°C. Concentrations and times are given in Table 5.2.
3. Wash roots three times, each time in 25 mL Tris buffer for 5 min each at 5°C.
4. Excise root meristems and transfer them to a 5 mL polystyrene tube containing 1 mL LB01 lysis buffer (in case of chromosome sorting for fluorescence in situ hybridization (FISH) and DNA amplification), 1.5 × IB buffer (in the case of chromosome sorting for HMW DNA preparation) or LB01-P buffer (in the case of chromosome sorting for proteomic analyses).

5. Isolate chromosomes by homogenizing. For appropriate settings (rpm, time) see Table 5.1.
6. Filter the suspension through 50 μm nylon mesh into a 5 mL polystyrene tube.
7. Store the suspension on ice.

   Although the chromosome suspension can be stored overnight, it is recommended to analyze the chromosomes on the same day.

   Alternatively, root tips may be homogenized using a razor blade. Transfer fixed root tips into 1.25 mL LB01 lysis buffer in a 6 cm glass petri dish. Chop meristem root tips individually using a sharp razor blade, avoiding dispersion or drying. Filter the suspension through a 50 μm nylon mesh into a polystyrene tube. Pass the suspension once through a 22-G needle to disperse intact metaphases, and store the suspension on ice. This method is more laborious and inconvenient in species with small root tips.

   However, it results in a higher yield of longer chromosomes in species with large or fragile chromosomes, such as field bean (see Table 5.2).

5.7.4   EXAMINE QUALITY OF CHROMOSOMES

1. Transfer 50 μL chromosome suspension into a 0.5 mL PCR tube.
2. Add 1 μL of 0.1 mg/mL DAPI stock solution.
3. Place a small drop (~10 μl) of DAPI-stained suspension on a microscope slide.
4. Using a fluorescence microscope, observe the suspension under low magnification (10× to 20×). Do not cover with a coverslip.

   The suspension should contain intact nuclei and chromosomes. The concentration of chromosomes in the sample should be ≥5 × 10^5/mL. If the chromosomes are damaged (broken and/or appear as long extended fibers), the formaldehyde fixation was too weak and should be prolonged. If the chromosomes are aggregated and/or the cells remain intact, the fixation was too strong and should be shortened.

5.8   BASIC PROTOCOL 3

5.8.1   UNIVARIATE FLOW KARYOTYPING OF PLANT CHROMOSOMES

This protocol describes the analysis and sorting of intact plant chromosomes stained with DAPI. The flow cytometer must be equipped with a UV light source to excite this dye.

5.8.2   MATERIALS

- Chromosome suspension (see basic protocol 2)
- 0.1 mg/mL DAPI stock solution (see recipe)
- LB01 buffer (see recipe)
- Collection liquid

- Sheath fluid for flow cytometric analysis (SF50): 50 mM NaCl in dH_2O (sterilize by autoclaving)
- 20 μm pore size nylon mesh in 4×4 cm^2
- Flow cytometer and sorter (e.g., FACSAria, BD Biosciences) with a UV laser (e.g., Coherent Genesis) and appropriate optical bandpass filter (e.g., 450 ± 25 nm)
- Additional reagents and equipment for aligning the flow cytometer and adjusting the sorting device (see support protocol 2), and for determining the purity of sorted chromosomes (see support protocol 3)

5.8.3 PERFORM FLOW CYTOMETRY

1. Filter the suspension through a 20 μm nylon mesh.
2. Stain chromosome suspension (1 mL/sample) by adding 0.1 mg/mL DAPI stock solution to a final concentration 2 μg/mL.
   Analysis can be performed immediately after addition of DAPI, without incubation. If necessary, the stained suspension can be kept on ice.
3. Make sure that the flow cytometer is properly aligned for univariate analysis (see support protocol 2) and that an appropriate optical filter is placed in front of the DAPI fluorescence detector.
4. Run a dummy sample (LB01 lysis buffer containing 2 μg/mL DAPI) for several minutes to equilibrate the sample line.
   This step is optional but advisable as it ensures stable peak positions during chromosome analysis and sorting.
5. Introduce the sample and let it stabilize at the appropriate flow rate (typically 1000–2000 particles/s). If possible, do not change the flow rate during the analysis.
   Significant changes in the flow rate during the analysis may result in peak shifts.
6. Set a gating region on a dot plot of forward scatter (FSC) and DAPI peak/pulse height to exclude debris, nuclei, and large clumps. Apply this gate to other histograms and dot plots.
7. Adjust photomultiplier amplification so that chromosome peaks are evenly distributed on a histogram of DAPI signal pulse area/integral.
8. Collect 20,000–50,000 chromosomes and save the results on a computer disk.

5.8.4 SORT CHROMOSOMES

1. Make sure that the sorting device is properly adjusted (see support protocol 2).
2. Run the sample and display the signals on a dot plot of DAPI signal pulse width versus area/integral.
3. Check for stability of the break-off point and of the side streams. Set appropriate level of deflection of side streams using plate voltage adjustment so that sorted chromosomes reach the collection vessel.

4. Define sorting region for chromosome of interest on the dot plot of DAPI pulse width versus DAPI pulse area/integral.
5. Select sort precision mode according to required purity and yield. Consult manufacturer's instructions for explanation of sort precision modes.
6. Sort the required number of chromosomes into a collection vessel containing appropriate amount of collection liquid.
   The amount and composition of the collection liquid depends on the number of sorted chromosomes and on their subsequent use.
7. Sort chromosomes onto a microscope slide for determination of purity (see support protocol 3).

5.9   ALTERNATE PROTOCOL 4

5.9.1   Bivariate Flow Karyotyping of Plant Chromosomes after FISHIS

This protocol is based on the method developed by Giorgi et al. (2013). The protocol gives a procedure for bivariate analysis and chromosome sorting after staining with DAPI and labeling specific sequences on chromosomes using fluorescence *in situ* hybridization in suspension (FISHIS).

5.9.2   Materials

- Chromosome suspension (see basic protocol 2)
- Ice
- 10M NaOH (see recipe)
- 1 M Tris·Cl (see recipe)
- Oligonucleotide microsatellite probe (e.g., $(GAA)_7$ labeled with FITC [Sigma-Aldrich])
- DAPI stock solution
- 20 μm pore size nylon mesh in 4×4 cm^2 pH meter
- Flow cytometer and sorter (e.g., FACSAria, BD Biosciences) with UV laser and optical path for DAPI excitation and detection (see basic protocol 3) and blue laser (488 nm; e.g., Coherent Sapphire) for FITC excitation, and appropriate optical bandpass filter (e.g., 530 ± 15 nm) for FITC detection.

5.9.3   Label DNA Repeats

A. Use 1 mL of chromosome suspension prepared in basic protocol 2. Keep on ice.
   As the following procedure dilutes the sample, use ~1.5× more root tips for sample preparation than usual.
B. Add 10 M NaOH to reach pH in the range 12.8–13.3.
C. Incubate the suspension for 15 min on ice.
D. Adjust the pH in the range of 8.5–9.1 using Tris·Cl.

E. Add (GAA)$_7$ probe working solution (1 ng/μL) to final concentration 180 ng/
mL and let the suspension incubate for 1 h in the dark at room temperature.
F. Put the suspension on ice and keep until the flow cytometric analysis.

5.9.4 Perform Flow Cytometry

A. Filter the sample through a 20 μm nylon mesh.
B. Stain the chromosome suspension by adding 0.1 mg/mL DAPI stock solution to a final concentration of 2 μg/mL.
C. Introduce the sample and let it stabilize at the appropriate flow rate (typically 1000–2000 particles/s). If possible, do not change the flow rate during
the analysis.
*Significant changes in the flow rate during the analysis may result in peak
shifts.*
D. Set a gating region on a dot plot of forward scatter (FSC) and DAPI peak/
pulse height to exclude debris, nuclei, and large clumps. Apply this gate to
other histograms and dot plots.
E. Adjust photomultiplier amplification of DAPI fluorescence detector so that
chromosome peaks are evenly distributed on a histogram of DAPI signal
pulse area/integral. Adjust photomultiplier voltage for FITC fluorescence
detector so that chromosome populations are evenly distributed on the
scale. Choose either linear or logarithmic amplification, depending on the
range of fluorescence signal intensities of individual populations.
F. Display the data on a dot plot of DAPI pulse area/integral versus FITC
pulse area/integral or FITC log.
G. Collect 20,000–50,000 chromosomes and save the results on a computer disk.

5.9.5 Sort Chromosomes

A. Make sure that the sorting device is properly adjusted (see support protocol 2).
B. Check for stability of the break-off point and of the side streams.
C. Define sorting region for the chromosome of interest on the dot plot of
DAPI pulse area/integral versus FITC pulse area/integral or FITC log.
D. Proceed with cell sorting as described (see basic protocol 3, steps 13–15).

5.10 SUPPORT PROTOCOL 2

5.10.1 Preparation of Flow Sorter for Chromosome Sorting

The alignment of the flow cytometer is crucial to achieve the highest purity in the
sorted chromosome fraction. This protocol describes suitable setups and fine tuning
of the instrument for chromosome analysis and sorting. This protocol is optimized
for use with BD FACSAria flow cytometer (BD Biosciences). For use with other flow
cytometers, refer to appropriate manufacturer's instructions for operation and basic
alignment of the instrument.

5.10.2   ADDITIONAL MATERIALS (*ALSO SEE BASIC PROTOCOL 3*)

- Calibration beads for checking cytometer performance: for example, BD FACSDiva CS&T beads (BD Biosciences)
- Calibration beads for sorting setup: for example, Accudrop beads (BD Biosciences) (optional); commercial fluorescent calibration beads, for example, Align Flow from Molecular Probes or Rainbow from Spherotec

5.10.3   SET UP FLOW CYTOMETER

1. Empty the waste container and fill the sheath container with sterile sheath fluid SF50.
2. Turn on the cytometer main power. Start up the workstation. Start BD FACSDiva software.
3. Perform fluidics startup.
4. Install a nozzle (70 μm orifice) and check for air bubbles (perform debubble after tank refill procedure, if necessary).
5. Select appropriate cytometer configuration (e.g., 70 μm orifice, 70 psi, blue laser detection path, UV laser detection path).
6. Start the stream.
7. Switch on the laser(s) and allow the laser(s) stabilize for ~10 min.
8. Perform cytometer performance check using appropriate beads (BD FACSDiva CS&T beads). If necessary, align the optical path of laser beams using fluorescent calibration beads to achieve maximum signal intensity and minimum coefficient of variation of DAPI and FITC signals (follow manufacturer´s instructions).

5.10.4   ADJUST SORTING DEVICE

1. Load a tube filled with a suspension of BD Accudrop beads and adjust the event rate of 1000–3000 events per second.
2. Turn on the voltage of deflection plates and start sorting.
3. Click the optical filter button on.
4. Optimize the drop delay by adjusting the drop delay value in 0.03-drop increments to achieve close to 100% intensity of the sorted stream of beads.
5. Click the optical filter button off.
6. Turn off the voltage of deflection plates.
7. Unload the tube of Accudrop beads.

5.11   BASIC PROTOCOL 4

5.11.1   CHROMOSOME SORTING FOR DNA AMPLIFICATION

This protocol is based on the method developed by Šimkové et al. (2008a). The protocol describes multiple displacement amplification of DNA from flow-sorted chromosomes. Amount, quality, and representativeness of DNA produced according to this protocol are suitable for sequencing by NGS technologies.

5.11.2 MATERIALS

- Deionized water (dH$_2$O)
- Chromosome suspension (see basic protocol 2)
- Proteinase K buffer 40× (see recipe)
- Proteinase K stock solution (10 mg/mL; see recipe)
- Illustra GenomiPhi V2 DNA amplification kit (GE Healthcare)
- 0.5 mL PCR tubes (sterilized by autoclaving)
- Microcentrifuge
- PCR cycler with heated lid
- Vivacon 500 columns, MWCO 100,000 Dalton (Sartorius)
- 1.5 mL PCR tubes
- Fluorimeter

5.11.3 SORT CHROMOSOMES

1. Prepare 0.5 mL PCR tubes containing 40 µL sterile dH$_2$O.
2. Sort appropriate number of chromosomes corresponding to 50 ng DNA. Typically, $2^{11} \times 10^4$ chromosomes are sorted depending on their molecular size (DNA content). Spin down briefly using a microcentrifuge to collect the chromosomes on the bottom of the tube.
3. If not processed immediately, sorted chromosomes can be stored up to 6 months at −20°C.

5.11.4 PURIFY AND AMPLIFY DNA

1. Thaw the chromosome fraction at room temperature.
2. Add proteinase K stock solution and 40× proteinase buffer so that the final concentration of proteinase K reaches 60 ng/µL and that of proteinase buffer reaches 1×. Incubate for 20 h at 50°C on a PCR cycler.
3. Add half the amount of proteinase K stock solution used in Step 5 and incubate another 20 h at 50°C on a PCR cycler.
4. To remove proteinase K and buffer, use a Vivacon 500 column. Add dH$_2$O to the sample in the column to reach 500 µL volume and centrifuge for approximately 10 min at 3000 × g, 24°C.
5. Repeat step 7 three times. Do not let the sample dry. The remaining volume should be about 10 µL.
6. Turn the column bottom-up and transfer the sample into a 1.5 mL PCR tube by centrifuging for 2 min at 2500 × g, 24°C.
7. Estimate the concentration of purified DNA using a fluorimeter.
   Note: *The yield reaches typically about 50% of the original DNA amount.*
8. Use 10 ng of purified DNA as a template for amplification.
9. Reduce the volume to 1 µL by overnight evaporation at 4°C and/or short evaporation at 50°C.
10. Amplify the DNA using Illustra GenomiPhi V2 DNA amplification kit. Follow the manufacturer's instructions.
11. Store up to 1 year at −20°C.

5.12 ALTERNATE PROTOCOL 5

5.12.1 SINGLE CHROMOSOME SORTING AND AMPLIFICATION

This protocol is based on the method developed by Cápal et al. (2015). The protocol outlines a procedure for production of micrograms of DNA from single copies of flow-sorted plant chromosomes.

5.12.2 MATERIALS

- Illustra GenomiPhi V2 DNA amplification kit (GE Healthcare)
- Proteinase K (10 mg/mL; Sigma-Aldrich, Cat. no. P4850)
- Lysis buffer (see recipe)
- Neutralization buffer (see recipe)
- Ice
- 0.2 mL PCR tubes (sterilized by autoclaving)
- Microcentrifuge (e.g., MiniStar silverline, VWR)
- PCR cycler with heated lid
- Sterile filter pipette tips (e.g., Axygen, Corning)

5.12.3 SORT CHROMOSOMES

1. Prepare 0.2 mL PCR tubes containing 3 µL mix of GenomiPhi V2 sample buffer and proteinase K (ratio 10:1).
2. Sort one chromosome into each tube. Use "single cell" sorting mode. For a positive control, sort 1000 chromosomes from the same sorting region into another tube. As a negative control, use a tube containing only the reaction mix.
3. Spin down the tubes using a microcentrifuge.

5.12.4 AMPLIFY DNA

1. Incubate the samples overnight at 50°C.
2. Inactivate the proteinase K by heating at 85°C for 15 min in a PCR cycler.
3. Store the samples up to 6 months at −20°C until use.
4. Thaw the samples at room temperature.
5. Add 1.5 µL lysis buffer and incubate for 15 min at 30°C. All pipetting steps should be performed in a UV-irradiated biohazard cabinet using sterile filter pipette tips.
6. Add 1.5 µL neutralization buffer. Spin down. Keep on ice until use.
7. Add DNA amplification master mix (4 µL sample buffer, 9 µL reaction buffer, and 1 µL enzyme; GenomiPhi V2 DNA amplification kit).
8. Incubate for 4 h at 30°C.
9. To stop the amplification process, incubate for 10 min at 65°C.

5.13 ALTERNATE PROTOCOL 6

5.13.1 CHROMOSOME SORTING FOR PREPARATION OF HIGH MOLECULAR WEIGHT DNA (HMW DNA)

This protocol is based on the method developed by Šimková et al. (2003). The protocol gives a procedure for preparation of HMW DNA suitable for construction of bacterial artificial chromosome (BAC) libraries and optical mapping using flow-sorted chromosomes.

5.13.2 MATERIALS

- 1.5× isolation buffer (1.5× IB; see recipe)
- 2% low-gelling agarose (InCert Agarose, Lonza) in 1× IB without
- 2-mercaptoethanol
- Lysis buffer B (see recipe)
- Lysis buffer C (see recipe)
- ET buffer (see recipe)
- Proteinase K
- 1.5 mL polystyrene tubes with conical bottom
- Centrifuge with swinging-bucket rotor
- Vortex mixer
- 50°C water bath
- Plug mold (Bio-Rad)

5.13.3 PREPARE HMW DNA FROM FLOW-SORTED CHROMOSOMES

1. For BAC library construction, sort chromosomes in batches of 200,000 into 1.5 mL polystyrene cuvettes containing 220 µL of 1.5× IB. For optical mapping, sort chromosomes in batches of 700,000 into 1.5 mL polystyrene cuvettes containing 770 µL of 1.5× IB. Number of batches depends on the DNA content of particular chromosome and the amount of HMW DNA needed.
2. Pellet the chromosomes by centrifuging in swinging-bucket rotor for 30 min at $300 \times g$, 4°C with slow breaking. Avoid any shaking before supernatant removal.
3. Carefully remove all of the supernatant except for 15 µL. Vortex gently for 5 s and prewarm to 50°C in a water bath.
4. Add 8 µL melted 2% agarose in 1× IB without mercaptoethanol preheated to 50°C. Vortex gently for 5 s and incubate in a water bath for 5 min at 50°C.
5. Pipette the chromosomes embedded in agarose with a cut off tip into a prewarmed plug mold. Let it solidify at 4°C for 5–10 min.
6. Incubate the agarose miniplugs with gentle shaking for 24 h at 37°C in lysis buffer C supplemented with freshly prepared proteinase K (0.1 mg/mL final concentration).

7. Exchange for the same amount of lysis buffer B containing proteinase K (0.1 mg/mL final concentration) and incubate for another 24 h at 37°C.
8. Wash the miniplugs with several milliliters ET buffer and store in ET buffer for up to 6 months at 4°C.

5.14 ALTERNATE PROTOCOL 7

5.14.1 CHROMOSOME SORTING FOR PROTEOMIC ANALYSES

This protocol is based on the method developed by Petrovská et al. (2014). The protocol describes a procedure for flow cytometric isolation of intact plant chromosomes suitable for high-throughput identification of chromosomal proteins using mass spectrometry.

5.14.2 MATERIALS

- Phenylmethanesulfonylfluoride (PMSF) stock solution (see recipe)
- LB01-P buffer (see recipe)
- 15 mL tubes with conical bottom
- Centrifuge

5.14.3 METHODS

1. Prepare 15 mL tubes containing 1 mL LB01-P supplemented with 150 μL PMSF stock solution.
2. Sort 10,000,000 chromosomes.
3. Pellet the chromosomes by centrifugation in swinging-bucket rotor for 30 min at $1000 \times g$, 4°C, with slow breaking.
4. Remove the supernatant. Store up to 1 year at −20°C until further processing.
5. For protein extraction, electrophoretic separation and mass spectroscopy analysis follow the protocol of Petrovská et al. (2014).

5.15 SUPPORT PROTOCOL 3

5.15.1 ESTIMATION OF PURITY IN SORTED FRACTIONS USING FISH

During the initial flow karyotyping experiment in a given plant species or line, it is useful to sort chromosomes onto a microscopic slide to determine the chromosome content of individual populations on a flow karyotype, and to examine the purity that may be achieved when sorting specific chromosomes. Each time chromosomes are sorted for subsequent analyses (see basic protocol 4 and alternate protocols 5–7), chromosomes should be also sorted onto a microscopic slide. This permits examination of the purity of a sorted chromosome fraction achieved under the actual sorting conditions.

This protocol allows precise estimation of the purity of a sorted fraction. The procedure is based on fluorescent labeling repetitive DNA sequences, which provide chromosome-specific patterns and thus enables discrimination of individual chromosomes. The probes are prepared using PCR with labeled nucleotide, hybridized to sorted chromosomes using FISH and signals are evaluated using a fluorescence microscope.

5.15.2 MATERIALS

- Deionized water (dH$_2$O)
- P5 buffer (see recipe)
- Chromosome suspension (see base protocol 2)
- Hybridization mix (see recipe)
- Rubber cement
- 2× SSC washing buffer (see recipe)
- 0.1× SSC stringent washing buffer (see recipe)
- 4× SSC washing buffer (see recipe)
- Blocking buffer (Amersham Biosciences)
- Fluorescently or hapten-labeled DNA probes
- Vectashield DAPI antifade solution (Vector Laboratories)
- Antidigoxigen-FITC (Roche)
- Streptavidin-CY3 (Invitrogen)
- Microscopic slides and coverslips
- PCR cycler with heated lid and *in situ* hybridization adapter
- Humidity chamber (37°C)
- Tweezers
- Parafilm
- Coplin staining jars
- Heated water bath with controlled temperature
- Fluorescence microscope with epi-illumination and filter sets for selected fluorochromes

5.15.3 SORT CHROMOSOMES

1. Pipette 7 μL P5 buffer onto a clean microscopic slide.
2. Sort ~2000 chromosomes into the drop.
3. Air dry for 3 h.
4. *Optional:* Keep overnight in the dark at room temperature until use.

5.15.4 PERFORM FISH REACTION

1. Add 25 μL hybridization mix, place a coverslip, and seal with rubber cement.
2. Let the DNA of chromosomes denature for 40 s at 80°C on a PCR cycler.
3. Put the slide into a humidity chamber and incubate overnight at 37°C.
4. Transfer the slide into a Coplin jar filled with preheated (42°C) 2× SSC and carefully remove the coverslip using tweezers. Wash in 2× SSC in a Coplin jar for 10 min at 42°C.
5. Wash in 0.1× SSC in a Coplin jar for 5 min at 42°C.

6. Wash in 2× SSC in a Coplin jar for 10 min at 42°C.
7. Remove the Coplin jar from the water bath. Replace the solution with heated (42°C) 2× SSC solution, and incubate the slide for 10 min at room temperature.
8. Wash in 4× SSC in a Coplin jar for 10 min at room temperature.
9. Remove the slide from the Coplin jar and add 60 μL of 1% blocking solution. Cover the slide with Parafilm and incubate for 10 min at room temperature.
10. *Optional:* Add fluorescently labeled antibody or streptavidin (follow manufacturer´s instructions regarding concentration) in 60 μL of 1% blocking solution, and incubate for 1 h at 37°C. Omit this step in the case of using fluorescently labeled DNA probes.
11. Wash the slide three times, each time with preheated (42°C) 4× SSC solution in a Coplin jar for 5 min at 42°C.
12. Rinse the slides (in dH$_2$O) in a Coplin jar to remove residual salt crystals and let them dry in the dark for at least 1 h.
13. Add Vectashield solution containing DAPI and cover with a coverslip.
    The amount of Vectashield depends on the coverslip size.
    R. Analyze the prepared slide using a fluorescence microscope. Examine with a DAPI filter first to localize the sorted chromosomes. For purity check of sorted chromosomes, evaluate at least 100 chromosomes and make statistics from three independent slides of the same-sorted chromosome.
    Avoid prolonged exposure to excitation light, which rapidly bleaches the fluorochromes.

5.15.5   REAGENTS AND SOLUTIONS

Use deionized water in all recipes and protocol steps.

5.15.5.1   Amiprophos-methyl (APM) Treatment Solutions
- *Stock solution*: Dissolve 0.0608 g APM (Ducheva) in 10 mL ice-cold acetone.
- Divide into 1 mL aliquots. Store up to 1 year at −20°C.
- *Working solution (2.5 μM)*: 101.3 μL APM stock solution in 800 mL dH$_2$O. Prepare just before use.
- *Working solution (5 μM)*: 202.6 μL APM stock solution in 800 mL dH$_2$O. Prepare just before use.
- *Working solution (10 μM)*: 405.2 μL APM stock solution in 800 mL dH$_2$O. Prepare just before use.

5.15.5.2   Blocking Reagent
- Dissolve 0.5 g blocking reagent (Sigma-Aldrich) in 50 mL of 4× SSC washing buffer by stirring for 1 h at 70°C. Sterilize by autoclaving and store in 1 mL aliquots up to 6 months at −20°C.

5.15.5.3   DAPI (4′, 6-Diamidino-2-phenylindole) Stock Solution, 0.1 mg/mL
- Dissolve 5 mg DAPI (Invitrogen) in 50 mL dH$_2$O by stirring for 60 min at room temperature. Pass through a 0.22-μm filter to remove small particles. Divide into 0.5 mL aliquots. Store up to 1 year at −20°C.

CAUTION: *DAPI is a possible carcinogen. It may be harmful if inhaled, swallowed, or absorbed through the skin, and may also cause irritation. Use gloves when handling. Be careful of particulate dust when weighing out the dye. Consult local institutional safety officer for specific handling and disposal procedures.*

5.15.5.4   ET Buffer

- 12.1 mg Tris base
- 1.862 g Na_2EDTA
- Adjust volume to 100 mL with dH_2O.
- Adjust final pH to 8.0 using 1N HCl.
- Sterilize by autoclaving.
- Store up to 6 months at 4°C.

5.15.5.5   Formaldehyde Fixative

- 0.303 g Tris base
- 0.931 g Na_2EDTA
- 1.461 g NaCl
- 250 µL Triton X-100 (0.1% [v/v] final)
- Adjust volume to 200 mL with dH_2O.
- Adjust pH to 7.5 using 1N NaOH.
- Add 37% (v/v) formaldehyde stock solution (Merck) as follows: 13.5 mL (2% solution), 20.25 mL (3% solution), or 27 mL (4% solution).
- Adjust final volume to 250 mL with H_2O.
- Prepare just before use.

CAUTION: *Formaldehyde is toxic and is also a carcinogen. It is readily absorbed through the skin and is irritating or destructive to the skin, eyes, mucous membranes, and upper respiratory tract. Wear gloves and safety glasses. Always work in a chemical fume hood. Consult local institutional safety officer for specific handling and disposal procedures.*

5.15.5.6   Fructose Syrup

- 30 g fructose
- 20 mL dH_2O
- Incubate overnight at 37°C.
- Add one crystal of thymol.
- Store up to 1 year at 4°C.

5.15.5.7   Hoagland's Nutrient Solution, 0.1×

- 10 mL of 10× stock solution
- 0.5 mL solution C (see recipe)
- Adjust volume to 1 L with dH_2O
  Prepare just before use.
  This recipe is from Gamborg and Wetter (1975).

5.15.5.8 Hoagland's Nutrient Solution, 1×
- 100 mL 10× stock solution
- 5 mL solution C (see recipe)
- Adjust volume to 1 L with dH_2O.
  Prepare just before use.

5.15.5.9 Hoagland's Solution A
- 280 mg H_3BO_3
- 340 mg $MnSO_4 \cdot H_2O$
- 10 mg $CuSO_4 \cdot 5H_2O$
- 22 mg $ZnSO_4 \cdot 7H_2O$
- 10 mg $(NH4)6Mo7O24 \cdot 4H_2O$
- Adjust volume to 100 ml with dH_2O
- Store up to 1 year at 4°C.

5.15.5.10 Hoagland's Solution B
- 0.5 mL concentrated H_2SO_4
- Adjust volume to 100 mL with dH_2O.
  Store up to 1 year at 4°C.

5.15.5.11 Hoagland's Solution C
- 3.36 g Na_2EDTA
- 2.79 g $FeSO_4 \cdot 7H_2O$
- Adjust volume to 400 mL with dH_2O.
- Heat to 70°C while stirring until the color turns yellow-brown.
- Cool and then adjust volume to 500 mL with dH_2O.
- Store up to 1 year at 4°C.

5.15.5.12 Hoagland's Stock Solution, 10×
- 4.7 g Ca $(NO3)2 \cdot 4H_2O$
- 2.6 g $MgSO4 \cdot 7H_2O$
- 3.3 g KNO3
- 0.6 g $(NH4) H_2PO_4$
- 5 mL solution A (see recipe)
- 0.5 mL solution B (see recipe)
- Adjust volume to 500 mL with dH_2O.
- Store up to 1 year at 4°C.

5.15.5.13 Hybridization Mix (FISH)
- 10 μL formamide, deionized, nuclease and proteinase free (final concentration 40%)
- 1.25 μL of 20× SSC
- 0.625 μL calf thymus DNA (10 mg/mL; final concentration 250 ng/μL)
- Labeled DNA probe(s) (20–200 ng/slide)
- Add 50% (v/v) dextrane sulfate to a final volume 25 μL.
  Prepare just before use.

5.15.5.14   Hydroxyurea Treatment Solutions

- *Working solution (0.5 mM)*: Dissolve 0.0304 g hydroxyurea (Sigma-Aldrich) in 800 mL Hoagland´s nutrient solution (see recipe). Prepare just before use.
- *Working solution (1.0 mM)*: Dissolve 0.0608 g hydroxyurea (Sigma-Aldrich) in 800 mL Hoagland's nutrient solution (see recipe). Prepare just before use.
- *Working solution (1.25 mM)*: Dissolve 0.0760 g hydroxyurea in 800 mL Hoagland's nutrient solution (see recipe). Prepare just before use.
- *Working solution (1.5 mM)*: Dissolve 0.0912 g hydroxyurea in 800 mL Hoagland's nutrient solution (see recipe). Prepare just before use.
- *Working solution (2 mM)*: Dissolve 0.1216 g hydroxyurea in 800 mL Hoagland's nutrient solution (see recipe). Prepare just before use.
- *Working solution (3 mM)*: Dissolve 0.1824 g hydroxyurea in 800 mL Hoagland's nutrient solution (see recipe). Prepare just before use.
- *Working solution (4.5 mM)*: Dissolve 0.2736 g hydroxyurea in 800 mL Hoagland's nutrient solution (see recipe). Prepare just before use.

5.15.5.15   Isolation Buffer, 1.5× (1.5× IB)

- 1.5 mL 10× IB (see recipe)
- 11 μL 2-mercaptoethanol
  Adjust volume to 10 mL with dH$_2$O.
  Add mercaptoethanol just before use.

5.15.5.16   Isolation Buffer, 10× (10× IB)

- 606 mg Tris base
- 1.86 g Na$_2$EDTA
- 174 mg spermine·4HCl
- 127.3 mg spermidine·3HCl
- 4.84 g KCl
- 584 mg NaCl
- 500 μL Triton X-100
  Adjust volume to 50 mL with dH$_2$O.
  Adjust final pH to 9.4 using NaOH pellets.
  Sterilize by filtration using a 0.22 μm filter.
  Store up to 3 months at 4°C.
  This recipe is from Šimková et al. (2003).

5.15.5.17   LB01 Buffer

- 0.363 g Tris base
- 0.149 g Na$_2$EDTA
- 0.035 g spermine·4HCl
- 1.193 g KCl
- 0.234 g NaCl
- 200 μL Triton X-100
  Adjust volume to 200 mL with dH$_2$O.
  Adjust final pH to 9 using 1N HCl.
  Filter through a 0.22-μm filter.

Add 220 µL 2-mercaptoethanol and mix well.
Divide into 10 mL aliquots.
Store up to 1 year at −20°C.
This recipe is from Doležel et al. (1989).
CAUTION: *2-Mercaptoethanol may be fatal if inhaled or absorbed through the skin and is harmful if swallowed. High concentrations are extremely destructive to the skin, eyes, mucous membranes, and upper respiratory tract. Wear gloves and safety glasses and work in a chemical fume hood.*

5.15.5.18   LB01-P Buffer, 1×

Mix 1 mL of the 10× LB01-P stock solution (see recipe) in 9 mL dH$_2$O. Add 10 µL of 2-mercaptoethanol and mix well. Prepare just before use.

5.15.5.19   LB01-P Buffer, 10×

- 1.817 g Tris base
- 0.744 g Na$_2$EDTA
- 0.190 g EGTA
- 0.070 g spermine·4HCl
- 0.127 spermidine·3HCl
- 5.96 g KCl
- 1.17 g NaCl
- 1000 µL Triton X-100
  Adjust volume to 100 mL with dH$_2$O.
  Adjust final pH to 8.0 using 1 M NaOH.
  Filter through a 0.22-µm filter.
  Divide into 10 mL aliquots.
  Store up to 1 year at −20°C.

5.15.5.20   Lysis Buffer (Single Chromosome Amplification)

- 0.280 g KOH
- 0.0154 g dithiothreitol
- 0.003722 g Na$_2$EDTA
  Adjust volume to 1 mL with dH$_2$O.
  Filter through 0.22 µm filter.
  Divide into 50 µL aliquots.
  Store up to 6 months at −20°C.

5.15.5.21   Lysis Buffer B (HMW DNA)

- 18.62 g Na$_2$EDTA
- 1 g sodium lauroyl sarcosine
  Adjust volume to 100 mL with dH$_2$O.
  Adjust final pH to 8.0 using NaOH pellets.
  Sterilize by autoclaving.
  Store up to 2 months at room temperature.

5.15.5.22　Lysis Buffer C (HMW DNA)

- 18.62 g Na$_2$EDTA
- 1 g sodium lauroylsarcosine
  Adjust volume to 100 mL with dH$_2$O
  Adjust final pH to 9.0 using NaOH pellets
  Sterilize by autoclaving
  Store up to 2 months at room temperature

5.15.5.23　NaOH Solution, 10 M

Dissolve 4 g solid NaOH in 7 mL dH$_2$O.
Adjust volume to 10 mL using dH$_2$O.
Store up to 2 months at room temperature.

5.15.5.24　Neutralization Buffer (Single Chromosome Amplification)

- Mix 53 µL concentrated HCl with 847 µL Tris·Cl, pH 8 (see recipe)
- Filter through a 0.22-µm filter.
- Divide into 50 µL aliquots.
- Store up to 6 months at −20°C.

5.15.5.25　Oryzalin Solutions

- *Stock solution*: Dissolve 0.0866 mg oryzalin (Ducheva) in 25 mL ice-cold acetone.
- Divide into 1 mL aliquots. Store up to 1 year at −20°C.
- *Working solution (2.5 µM)*: 200 µL oryzalin stock solution in 800 mL dH$_2$O.
- Prepare just before use.
- *Working solution (5 µM)*: 400 µL oryzalin stock solution in 800 mL dH$_2$O.
- Prepare just before use.
- *Working solution (10 µM)*: 800 µL oryzalin stock solution in 800 mL dH$_2$O.
- Prepare just before use.

5.15.5.26　P5 Buffer

- Dissolve the following in 25 mL dH$_2$O:
  30.28 mg Tris base
  93.2 mg KCl
  10.17 mg MgCl$_2$·6H$_2$O
  1.25 g sucrose
- Adjust pH to 8 using 1N HCl.
- Divide into 1 mL aliquots.
- Store up to 1 year at −20°C.

5.15.5.27　Phenylmethanesulfonylfluoride (PMSF) Stock Solution, 100 mM

- Dissolve 174.2 mg PMSF in 10 mL isopropanol.
- Divide into 150 µL aliquots.
- Store up to one year at −20°C.
- *Do not store thawed aliquots.*

5.15.5.28 Proteinase K Buffer, 40×
- 100 µL 1 M Tris·Cl, pH 8.0
- 100 µL 0.5M EDTA, pH 8.0
- 500 µL 10% SDS (Sigma-Aldrich)
- Adjust volume to 1 mL with sterile dH_2O.
- Store up to 3 months at room temperature.

5.15.5.29 Proteinase K Stock Solution, 10 mg/mL
- Dissolve 1 mg proteinase K (Roche) in 100 µL sterile dH_2O
- Store up to 1 week at 4°C.

5.15.5.30 Schiff's Reagent
- 30 ml 1 N HCl
- 2 g parafuchsin (Serva; color index 42500)
- 3.8 g $K_2S_2O_5$
- 170 mL dH_2O

Stir for 2 h in a tightly closed bottle and leave overnight. Add 2 g active charcoal, mix 1 min, and filter through a paper filter moistened with 1N HCl. Repeat filtration if solution is not colorless. Store up to 6 months in a tightly closed bottle at 4°C.

5.15.5.31 SSC Solutions
- *20× stock solution*: Dissolve 175.3 g NaCl and 88.2 g $Na_3C_6H_5O_7·2H_2O$ in dH_2O in final volume 1000 mL. Adjust pH to 7 using 1N HCl. Sterilize by autoclaving. Store for up to 3 months at room temperature.
- *4× SSC washing buffer*: Mix 20× SSC (200 mL) and 2 mL Tween 20. Add dH_2O up to 1000 mL. Sterilize by autoclaving.
- *2× SSC washing buffer*: 20× SSC (100 mL) in 900 mL dH_2O. Sterilize by autoclaving.
- *0.1× SSC stringent washing buffer*: 20× SSC (5 mL), 1 mL Tween 20, and 0.406 g 2 mM $MgCl_2·6H_2O$ and dH_2O up to 1000 mL. Sterilize by autoclaving.

5.15.5.32 Tris Buffer
- 0.606 g Tris base
- 1.861 g Na_2EDTA
- 2.922 g NaCl
- Adjust volume to 500 mL with dH_2O.
- Adjust pH to 7.5 using 1N NaOH.
- Store up to 6 months at 4°C.

5.15.5.33 Tris Cl, 1 M
- 60.5 g Tris base
- Dissolve Tris base in 400 mL dH_2O by stirring.

- Adjust the pH to 8 using 1N HCl.
- Adjust volume to 500 mL using dH_2O.
- Store up to 6 months at 4°C.

5.16 COMMENTARY

5.16.1 BACKGROUND INFORMATION

The majority of agriculturally important plant species have complex nuclear genomes which makes their analyses difficult. The daunting task may be simplified by dividing their genomes into well-defined parts—chromosomes. Generally, individual chromosomes can be isolated either by chromosome microdissection or by flow sorting. The advantage of flow cytometry is that chromosomes can be sorted in large numbers in a relatively short time. The salient feature of flow cytometry is that the target particles are suspended in a narrow stream of liquid (typically saline); they are forced to move in a single file, where they can be made to interact one-by-one with an orthogonally oriented light beam (Figure 5.2).

Flow analysis and sorting of mitotic chromosomes (flow cytogenetics) greatly stimulated progress in human and animal genome mapping. Chromosomes stained with DNA fluorochromes were individually classified according to their dye content, and the resulting distributions (flow karyotypes) were shown to be useful for detection of numerical and structural chromosome aberrations (Otto 1988; Cooke et al. 1989; Boschman et al. 1992). In the past, flow-sorted chromosomes were used mainly for gene mapping and integration of genetic and physical maps (Lebo 1982; Sargan et al. 2000), array painting (Fiegler et al. 2003; Veltman et al. 2003), construction of chromosome-specific DNA libraries (Van Dilla and Deaven 1990; Korstanje et al. 2001), and for generation of chromosome painting probes to study chromosome aberrations (Carter 1994; Sargan et al. 2005) or evolution and phylogeny of chromosomes (Ferguson-Smith and Trifonov 2007). When coupled to next-generation sequencing (NGS), flow-sorted chromosomes facilitate mapping chromosomal breakpoints (Chen et al. 2008, 2010) and allele phasing (Yang et al. 2011).

Attempts to develop flow cytogenetics in plants were hampered by difficulties in preparing suspensions of intact chromosomes and by the inability to discriminate single chromosome types (reviewed by Doležel et al. 1994). These problems were overcome by the development of new procedures and experimental approaches, and the progress of plant flow cytogenetics has accelerated in the recent years. While in 2001 Doležel et al. reported successful flow cytometric analyses in 12 plant species from 11 genera, Table 5.1 outlines the current status, with 32 species from 20 genera. In Table 5.2 we present protocols for 42 species from 25 genera developed in the Doležl laboratory, which includes unpublished protocols for 18 species. In addition, the range of applications of sorted chromosomes has grown from simple physical mapping of DNA sequences using FISH and PCR to sophisticated applications in genomics (reviewed in Doležel et al. 2012, 2014). The following paragraphs will deal with different aspects of plant flow cytogenetics, starting with the preparation of chromosome suspensions and ending with applications of sorted chromosomes, in more detail.

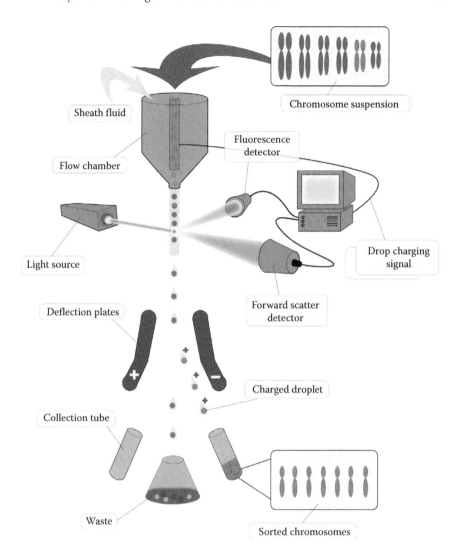

FIGURE 5.2 The mechanics of flow sorting. Chromosomes held in liquid suspension are stained by a fluorochrome and passed into a flow chamber containing a sheath fluid. The geometry of the chamber forces the chromosome suspension into a narrow stream in which the chromosomes become aligned in a single file, and so are able to interact individually with an orthogonally directed laser beam(s). Pulses of scattered light and emitted fluorescence are detected and converted to electric pulses. If the chromosome of interest differs in fluorescence intensity from other chromosomes, it is identified and sorted. The sorting is achieved by breaking the stream into droplets and by electrically charging droplets carrying chromosomes of interest. The droplets are deflected during their passage through the electrostatic field between defection plates and collected in suitable containers. (From Doležel et al. *Biotechnology Advances* 32, 122–136, 2017.)

5.16.2 MATERIALS USED FOR PREPARATION OF PLANT CHROMOSOME SUSPENSIONS

The first suspension of plant chromosomes suitable for flow cytometric analysis was prepared from suspension-cultured cells of slender goldenweed (De Laat and Blaas 1984). Subsequently, chromosomes were isolated from cultured cells of tomato and bread wheat (Arumuganathan et al. 1991, 1994; Wang et al. 1992). Cell suspensions grown *in vitro* appeared to be an attractive system for chromosome isolation mainly because large numbers of cells could be grown and manipulated under defined conditions. However, they have certain disadvantages, the most serious being their karyological instability. In the light of this fact, cell suspensions are no longer used for the preparation of chromosome suspensions for subsequent flow cytometric analyses.

Some authors isolated chromosomes from protoplasts derived from leaf cells (Conia et al. 1987, 1988, 1989). In the young leaf, the majority of mesophyll cells are reversibly arrested at the G0/G1 phase of the cell cycle. Upon transfer to the appropriate culture medium, protoplasts derived from these cells resume the cell cycle, traverse the S and G2 phases, and enter mitosis with a certain degree of synchrony. Although relatively simple, this approach has not been used frequently, mainly because the extent of relatively low synchrony, poor chromosome yields, and also because of difficulties with establishing protoplast cultures in many plant species.

The third system, developed by Doležel et al. (1992) and described in this chapter, involves the use of root-tip meristems and has been by far the most frequently employed. Examples of successful applications include agriculturally important legumes (Lucretti et al. 1993; Gualberti et al. 1996; Vláčilová et al. 2002) and cereals (Lee et al. 1996, 1997; Lysák et al. 1999; Vrána et al. 2000; Kubaláková et al. 2003). The advantage of roots is that they can be easily obtained from seeds for the majority of plant species, they are cheap to produce, and they are easy to grow and manipulate. A further advantage of root-tip meristems is their karyological stability. Although generally applicable, this experimental system may not be accessible for genotypes that are difficult to propagate via seeds. For this case we have developed a protocol for the preparation of chromosomes from roots grown from bulbs or stem cuttings. In the past, use of immortal (hairy root) cultures produced after infection with the bacterium *Agrobacterium rhizogenes* has been described (Veuskens et al. 1992, 1995; Neumann et al. 1998), however, due to their potential karyological instability, their further utilization is scarce.

5.16.3 CELL CYCLE SYNCHRONIZATION AND ACCUMULATION OF METAPHASE CHROMOSOMES

While "spontaneous" cell cycle synchrony is observed in cultured mesophyll protoplasts, it must be artificially induced in suspension-cultured cells and root tips. Most frequently, this has been achieved by treatment with DNA synthesis inhibitors like hydroxyurea or aphidicolin. The treatment leads to the accumulation of cycling cells mainly at the G1/S interface. After a suitable period, cells are released from the block and traverse the cycle synchronously. Although the mechanisms of action of hydroxyurea and aphidicolin are different (Young and Hodas 1964; Sala et al. 1980), the degree of induced cell cycle synchrony is similar. However, hydroxyurea is considerably cheaper and thus has been used more frequently.

In order to accumulate a sufficient number of synchronized cells in metaphase, chromosome movement in mitosis must be inhibited. This can be conveniently achieved using mitotic spindle poisons. One of them, the alkaloid colchicine, has been frequently employed in procedures for chromosome isolation (De Laat and Blaas 1984; Conia et al. 1987; Wang et al. 1992; Arumuganathan et al. 1994; Úberall et al. 2004). Although widely used, colchicine has disadvantages. Its affinity to plant tubulins is low, and thus it has been applied at relatively high (millimolar) concentrations. Another serious drawback of colchicine is its tendency to induce extensive chromosome stickiness and clumping. In contrast to colchicine, synthetic herbicides like amiprophos-methyl, oryzalin, and trifluralin show visible antimicrotubular effects at micromolar concentrations, and thus have been used frequently in chromosome isolation protocols (Doležel et al. 1992; Veuskens et al. 1995; Lee et al. 1996). In some species such as corn, treatment with mitotic spindle poisons induces chromosome clumping. A treatment with nitrous oxide has been used to avoid this negative effect (Kato 1999).

5.16.4 PREPARATION OF CHROMOSOME SUSPENSIONS

Chromosome isolation is hampered by the rigid cell walls typical of plant cells. Two approaches have been used to release chromosomes from synchronized cells. The first involves removal of the cell wall by the action of hydrolytic enzymes such as pectinases and cellulases. Protoplasts thus obtained are lysed in a hypotonic buffer, where they burst and liberate chromosomes (De Laat and Blaas 1984). The second approach, which is described in this chapter, involves direct mechanical release of chromosomes from synchronized root meristems after formaldehyde fixation (Doležel et al. 1992). Compared to the protoplast lysis method, the latter has several advantages. Mechanical isolation is rapid and avoids extended enzyme incubations, during which chromosome decondensation may occur. Moreover, enzyme mixtures used to digest cell walls may be contaminated by DNases.

Lee et al. (1996) introduced a modified version of Doležel's procedure where the fixation step was omitted. However, the original method based on formaldehyde fixation has several advantages. Fixed chromosomes are resistant to mechanical shearing, and thus can be isolated using a mechanical homogenizer, which takes only a few seconds. Mechanical stability of isolated chromosomes permits two-step sorting to improve the purity of sorted fractions (Lucretti et al. 1993). The fixation step also significantly increases the yield of chromosomes. The morphology of isolated chromosomes is well preserved and their DNA is suitable for PCR (Macas et al. 1993; Kejnovský et al. 2001). In addition, chromosomes prepared according to this protocol were shown to be suitable for scanning electron microscopy (Schubert et al. 1993), *in situ* hybridization (Fuchs et al. 1994; Vláčilová et al. 2002; Valárik et al. 2004; Molnár et al. 2011, 2014; Kopecký et al. 2013), primed *in situ*

DNA labeling (PRINS) (Kubaláková et al. 1997, 2000), immunolocalization of chromosomal proteins (Binarová et al. 1998), preparation of high molecular weight (HMW) DNA (Šimková et al. 2003), and linear amplification of DNA (Šimková et al. 2008a; Vrána et al. 2012).

5.16.5 FLOW KARYOTYPING

The majority of plant species have morphologically similar chromosomes with similar DNA content. Due to this, the number of chromosomes that can be discriminated and sorted in various species is usually low (Doležel et al. 1994, 1999). Different strategies have been employed to overcome this problem. Bivariate flow karyotyping is a standard method of human flow cytogenetics that allows discrimination of almost all chromosome types (Stap et al. 2001). With rare exceptions (Arumuganathan et al. 1994), this approach does not help to discriminate higher numbers of chromosomes in plants when compared to univariate analysis (Lucretti and Doležel 1997; Schwarzacher et al. 1997). The reason for this is most probably a high proportion of repetitive DNA sequences dispersed more or less evenly in plant genomes.

Lucretti et al. (1993) suggested the use of chromosome translocation lines to aid in discrimination and sorting of individual chromosome types. Although developed for the field bean (Doležel and Lucretti 1995), this approach has also been found useful for garden pea (Neumannn et al. 1998) and barley (Lysák et al. 1999), thus confirming its overall usefulness. Apart from resolving otherwise undiscriminable chromosomes, sorting from translocation or other lines with changed chromosome morphology such as deletions or alien chromosome additions (cytogenetic stocks) was found useful for localization of DNA sequences at subchromosomal level (Macas et al. 1993; Neumann et al. 2002), delimiting the translocations (Berkman et al. 2012) or mapping alien introgressions (Tiwari et al. 2014). The International Wheat Genome Sequencing Consortium (IWGS) adopted chromosome sorting from telosomic lines as primary strategy to produce chromosome based draft genome sequence of wheat (The International Wheat Genome Sequencing Consortium 2014).

If cytogenetic stocks are not available for the material of interest, a different approach exploiting labeling of specific DNA sequences might be favorable. The first method used in plant flow cytogenetic was primed in situ DNA labeling en suspension (PRINSES). This method was used to discriminate chromosomes of similar size in field bean (Macas et al. 1995; Pich et al. 1995). However, because the method was rather complicated and not generally reproducible, its use was limited. Recently, Giorgi et al. (2013) developed a method for labeling repetitive DNA sequences (especially microsatellites) on plant chromosomes in suspension (FISHIS). The method is reproducible and has been used for molecular and comparative analysis of chromosomes of several wild ancestors and relatives of wheat (Akpinar et al. 2015a; Molnár et al. 2016). A "low-tech" alternative to all the mentioned methods is chromosome sorting from composite peaks using narrow sorting windows. Although this approach does not have an ambition to provide pure chromosome fractions, it provides enriched fractions of chromosomes of interest and might be useful especially in polyploid species, where it could provide chromosomes of interest free of contaminating homeologs (Vrána et al. 2015). Another possibility to obtain chromosome-specific DNA, even if the chromosome of interest cannot be discriminated, is single chromosome sorting (Cápal et al. 2015). If only one copy of a chromosome is sorted, it is not contaminated by other chromosomes. However, as anonymous chromosomes are sorted, the identity of sorted chromosomes is revealed only after DNA amplification using PCR or after sequencing. Cápal et al. (2015) showed that sequencing three separate amplification products from single copies of wheat chromosome 3B resulted in 60% coverage of the chromosome with more than 30% of its genes fully covered. This approach could be utilized for any plant species, where discrimination of individual chromosomes is difficult and also to study chromosome structural heterozygosity and haplotype phasing.

As mentioned above, flow cytometry can detect chromosomal aberrations such as insertions, deletions, or translocations. Moreover, thanks to its high resolution, flow cytometry is capable to detect even small differences in DNA content. Vrána et al. (2015) identified decreased DNA content of chromosome 4A in some wheat cultivars. Tsõmbalova et al. (2016) in their detailed survey showed that this decrease in DNA content was associated with a specific haplotype in the distal part of chromosome 4A long arm (4AL).

5.16.6 Chromosome Sorting

The problems associated with discrimination of single chromosome types underline the importance of procedures for identification of the chromosomal content of individual peaks on a flow karyotype and for determining the purity of sorted chromosome fractions. While dot blotting (Arumuganathan et al. 1994) or PCR with primers for chromosome-specific markers (Lysák et al. 1999) may be sufficient for chromosome assignment, microscopic evaluation of a sorted fraction is needed to determine the frequency and the nature of contaminating particles.

Usually, morphological analysis of sorted chromosomes is not sufficient, as the length and even the arm ratio may differ significantly among sorted chromosomes of the same type. An elegant approach is to localize DNA sequences that are chromosome-specific or show chromosome-specific distribution. Thus, Lucretti et al. (1993) used FISH with a probe for rDNA to evaluate the purity of a sorted chromosome fraction from field bean. Over the years, FISH has become the universal method for purity estimation in many plant species, where chromosome sorting was conducted. An alternative to FISH is PRINS. This technique does not require labeled probes and is faster, thus permitting rapid evaluation of sorted chromosomes. Its usefulness to identify sorted chromosomes and evaluate their purity was demonstrated for pea (Gualberti et al. 1996), barley (Lysák et al. 1999), and wheat (Vrána et al. 2000).

5.16.7 Utility of Sorted Chromosome Fractions

Flow-sorted plant chromosomes were found invaluable in a number of studies. Table 5.3 lists the most popular applications, including the number of chromosomes and sorting time needed. Generation of chromosome-specific DNA libraries is one

TABLE 5.3
Number of Sorted Chromosomes and Sorting Times Needed for Different Applications

Application	Number of chromosomes	Sorting time
Single chromosome amplification	1	Seconds
Physical mapping using PCR	500–1,000	Tens of seconds
Physical mapping using FISH	1,000–2,000	Minutes
NGS sequencing of MDA-amplified chromosomes	25,000–110,000[a]	Hours
BAC libraries	200,000[a]	Weeks
Optical mapping	700,000[a]	Weeks
Proteomic analysis	10,000,000[a]	Weeks

[a] These numbers represent one batch. Total number of batches depends on molecular size of particular chromosome and DNA amount needed.

of the most important and earliest applications of sorted chromosomes. Such libraries (initially short-insert libraries) can significantly reduce the efforts required for isolation of DNA sequences including molecular markers from defined regions of the genome. Thus, Wang et al. (1992) constructed a chromosome-enriched DNA library from sorted bread wheat chromosome 4A. The library showed a 20-fold enrichment of clones as compared to a random genomic library. Arumuganathan et al. (1994) reported generation of a chromosome 2-specific DNA library in tomato. Using degenerate oligonucleotide-primed PCR (DOP-PCR), the library was constructed from only one thousand sorted chromosomes. Furthermore, the authors isolated 11 chromosome 2-specific probes that detected restriction fragment length polymorphisms (RFLPs) and placed them on the genetic linkage map of tomato. Macas et al. (1996) constructed a set of short-insert chromosome-specific DNA libraries covering the whole field bean genome. This was the first case in plants where a whole genome was available in the form of chromosome-specific DNA libraries. Although short-insert libraries were very useful (e.g., for gene mapping and development of DNA markers), the length of DNA inserts (~1–2 kb) did not enable more advanced applications such as construction of physical maps or positional gene cloning. This raised demand for chromosome-specific DNA libraries of large inserts. The first prerequisite for the construction of such libraries is a large amount of high molecular weight DNA (HMW DNA). A protocol for preparation of HMW DNA from flow-sorted plant chromosomes (Šimková et al. 2003) made production of microgram DNA amounts realistic, and Šafář et al. (2004) were able to construct a BAC library from the first plant chromosome (chromosome 3B of hexaploid wheat). More chromosome-specific BAC libraries soon followed suit (Janda et al. 2006; Šimková et al. 2008b; Šafář et al. 2010). The list of chromosome-specific BAC libraries prepared in our facility is available on our Web site (http://olomouc. ueb.cas.cz/genomic-resources).

Physical mapping of DNA sequences was among the first applications of flow-sorted plant chromosomes. The most prevalent techniques for physical localization of specific DNA sequences are PCR and FISH. Thanks to low number of chromosomes needed, PCR with sorted chromosomes was extensively used to localize specific DNA sequences at chromosomal (Kejnovský et al. 2001) or subchromosomal level (Macas et al. 1993; Lysák et al. 1999), and to integrate genetic and physical maps (Neumann et al. 2002; Vláčilová et al. 2002; Zatloukalová et al. 2011). More recently, Šimková et al. (2011) reported on successful PCR screening of chromosome-specific BAC libraries with markers for resistance genes. Using PCR on flow-sorted chromosomes, Cápal et al. (2016) identified barley transgene insertion sites in wheat. Lysák et al. (1999) and Vrána et al. (2000) demonstrated the utility of chromosomes sorted from barley and wheat, respectively, for physical mapping of microsatellite markers using PRINS. Over the years, FISH has proven itself as a very important cytogenetic tool for high-resolution mapping. FISH on flow-sorted chromosomes allowed to recognize individual chromosomes (Kubaláková et al. 2003; Suchánková et al. 2006; Grosso et al. 2012; Kopecký et al. 2013; Molnár et al. 2014), chromosome translocations or deletions (Kubaláková et al. 2003, 2005; Suchánková et al. 2006), presence of an alien chromosomes (Guo et al. 2006, 2013; Suchánková et al. 2006) and B chromosomes (Kubaláková et al. 2003), or localization of BAC clones on flow-sorted chromosomes (Šafář et al. 2004). Spatial resolution of *in situ* localizations of DNA sequences could be increased by longitudinally stretching the flow-sorted chromosomes while retaining their integrity. Valärik et al. (2004) improved spatial resolution down to 70 kbp as compared to 5–10 Mbp after FISH on mitotic chromosomes by stretching the sorted chromosomes with ethanol:acetic acid (3:1). Endo et al. (2014) showed that just by changing the conditions of chromosome suspension preparation (milder fixation) and chromosome sorting, one can obtain hyperexpanded chromosomes ~5 to 10 times longer than the same chromosomes from the squash preparations. The hyperexpanded chromosomes maintained their overall morphology and structure and were suitable for FISH.

DNA from flow-sorted chromosomes proved to be a suitable template for whole genome amplification using high-fidelity polymerases, such as Phi29 (Šimková et al. 2008a). This technique is capable of producing microgram quantities from starting amounts as low as 10 ng with a low amplification bias. The size of amplification products is typically in range of 5–20 Kb, which is suitable for the majority of NGS technologies. Coupling amplified DNA from flow-sorted plant chromosomes with NGS sequencing proved to be fast and affordable approach to obtain first insights into the genomes of species where reference sequences are not available yet. So far, sequencing DNA amplified from flow-sorted chromosomes has been reported in barley (Mayer et al. 2009, 2011), rye (Fluch et al. 2012; Martis et al. 2012, 2013), wheat (Wicker et al. 2011; Berkman et al. 2012; Hernandez et al. 2012), fescues (Kopecký et al. 2013), and goat grasses (Tiwari et al. 2015), and it helped to obtain novel information on gene content and their organization on particular chromosomes. Using flow-sorted chromosomes and NGS technology, Nie et al. (2012) were able to develop set of chromosome-arm-specific microsatellite markers in wheat. Recently, a large international effort resulted in an ordered draft genome sequence of the hexaploid bread wheat (Mayer et al. 2014). Sequencing flow-sorted chromosomes also helped to validate draft chickpea genome assemblies (Ruperao et al. 2014), and uncover the molecular composition of supernumerary B chromosomes in rye (Martis et al. 2012). It is expected that sequencing flow-sorted chromosomes of ancestors or wild relatives of important agronomical crops (e.g., wheat and barley) will help to develop new molecular tools to facilitate the identification of alien chromatin and support alien introgression breeding (Tiwari et al. 2014, 2015; Akpinar et al. 2015a, 2015b; Molnár et al. 2015, 2016). Recently, DNA prepared from flow-sorted plant chromosomes was found to be also suitable for optical mapping (Staňková et al. 2016). While the majority of uses of flow-sorted chromosomes describe the analysis of DNA, Petrovská et al. (2014) developed a protocol for the analysis of chromosomal proteins using mass spectrometry. This breakthrough will make it possible to characterize the proteome of plant mitotic chromosomes with the aim to reveal their organization at molecular level.

5.16.8 CRITICAL PARAMETERS AND TROUBLESHOOTING

5.16.8.1 Cell Cycle Synchronization and Metaphase

The quality of a chromosome suspension depends not only on the procedure for chromosome isolation, but also on the frequency of metaphase cells in the source tissue. Although the mitotic activity in meristem root tips may be relatively high, it is not sufficient for chromosome isolation. Thus, the protocols described here rely on artificial synchronization of the cell cycle based on the inhibition of cell cycle progression at the G1/S interface by the action of hydroxyurea. The effect of the treatment depends critically on the hydroxyurea concentration. Because cell cycle kinetics are sensitive to external factors including temperature, care should be taken to adjust the temperature of all solutions prior to use. Also, the length of all treatments should be monitored, and residual treatment solutions should be carefully washed away before transferring seedlings to other solutions. Seedlings with healthy and actively growing roots should always be used. To achieve reproducible results, seedlings with a similar root length should be selected. For most species with thick roots (e.g., legumes), optimal concentration of Hoagland´s nutrient solution is 1×, while 0.1× Hoagland´s nutrient solution is used in the case of cereals and grasses.

In the protocols presented in this unit, synchronized cells are arrested in metaphase by the action of mitotic spindle poisons, amiprophos-methyl, or oryzalin. Although higher frequencies of metaphase cells may be obtained after longer treatments, a short (typically 2 h) treatment is preferred as it decreases the proportion of single chromatids and avoids chromosome decondensation. Compared to other mitosis blockers, both amiprophos-methyl and oryzalin have a high affinity for plant tubulins and can be used in very low concentrations. If needed, other mitotic spindle inhibitors may be used instead. This will require determination of optimal concentrations. In some species, including cereals such as barley, wheat, and rye, overnight incubation of synchronized seedlings in ice water improves the spreading of chromosomes within cells. This, in turn, results in improved chromosome yield and lower frequency of chromosome clumps. In some species, where the action of mitotic poisons causes chromosome clumping, such as corn, nitrous oxide may be used (Kato 1999).

It is important to examine the frequency of metaphase cells in the root tips prior to chromosome isolation (see support protocol 1). If the frequency is <40%, or if the cells are not arrested in metaphase, check the procedure for metaphase accumulation. First, analyze the degree of mitotic synchrony induced by the hydroxyurea treatment. Take samples of root tips at 1 h intervals after removal from the hydroxyurea block and determine the mitotic activity on Feulgen-stained preparations. The mitotic activity should gradually increase and peak at ~50%. If this is not the case, check the quality of the plant material and of all reagents and solutions. Based on the authors' experience, treatment with a mitotic spindle inhibitor should be started 30–90 min before the peak of mitotic activity. If the synchronized cells are not arrested in metaphase, test the mitosis blocker using nonsynchronized cells (root tips). Try to determine the optimal concentration that will arrest cells at metaphase (no anaphase or telophase cells should be observed) and that will not induce chromosome clumping. The abovementioned trouble-shooting strategy may be also used to modify the procedure for other species not mentioned in this unit.

5.16.8.2 Preparation of Chromosome Suspension

Start with the formaldehyde fixation of root tips immediately after finishing the incubation in the mitosis blocker. The extent of formaldehyde fixation is a critical determinant of chromosome morphology and yield. Thus, the concentration of form-aldehyde, the temperature of the fixative, and the length of fixation must be strictly followed. In addition, note that the formaldehyde fixative used in this protocol is not neutralized. Check the quality of the chromosome suspension under a fluorescence microscope. Weak fixation results in poor preservation of chromosome morphology. Most of the isolated chromosomes are damaged (extended fibers) and the suspensions contain a large amount of chromosome debris. However, if the fixation is too strong, the homogenized suspensions will contain large numbers of chromosome clumps and even intact cells. When starting with the chromosome preparation protocol, it is recommended to test several variants of fixation. For instance, change the length of the fixation in 5 min steps, or change the concentration of formaldehyde in 1% steps, and observe the effect under the microscope. Always compare chromosome suspensions prepared on the same day from the same batch of synchronized root tips. Select the optimum fixation giving the highest yield of morphologically intact chromosomes.

However, the final decision on the optimum extent of fixation should be based on flow cytometric analysis of isolated chromosomes. The fluorescence distribution (flow karyotype) should contain well-resolved chromosome peaks with the least amount of background debris, chromatids, and chromosome clumps. Storage of fixed root tips in Tris buffer prior to chromosome isolation may decrease chromosome yield, so it is advisable to perform the isolation immediately after fixation. Although suspensions of isolated chromosomes may be stored at 4°C, the best resolution of flow karyotypes is usually obtained with freshly prepared samples.

5.16.8.3 Flow Karyotyping

The resolution of flow karyotypes depends critically on the performance of the cytometer. Always make sure that the instrument is properly aligned prior to starting chromosome analysis. This is usually done using suitable fluorescent microspheres to achieve the lowest coefficient of variation (CV) for fluorescent peaks. Naturally, the lowest CV that can be achieved depends also on the quality of the calibration beads. Use the best particles available with CV ≤1.5%. If the resolution of fluorescent peaks remains poor, check the fluidic system and the nozzle. All tubing must be clean and free of air bubbles. A dirty nozzle is probably the most frequent reason for poor resolution in jet-in-air flow sorters. When in doubt, always clean the nozzle. This can be done in a small sonicating water bath, but note that only the nozzle itself should be sonicated, as the holder may disintegrate. Similarly, a dirty flow cell in cuvette-based systems also impairs the resolution. In this case follow the manufacturer's instructions for cleaning procedures (e.g., "clean flow cell" procedure for BD FACSAria). A convenient way to check the state of the nozzle and fluidics is by analysis of forward light scatter. The resulting distribution should be tight and have a CV similar to that observed in fluorescence channels.

Compared to human or animal chromosome suspensions, plant chromosome suspensions are usually less concentrated, and therefore, should be run at lower speeds (e.g., 1000–2000 chromosomes/sec) to achieve the best resolution. The peak position can show instability due to slow equilibration of fluorescent dyes in the sample line. To minimize this problem, it is advisable to run a dummy sample prior to analysis of the suspension. Even then, it may take a few minutes before the chromosome peak positions are stable and data may be acquired. Because isolated plant chromosomes vary greatly in the degree of condensation and, hence, also in length, better resolution is obtained on histograms of fluorescence pulse area rather than fluorescence pulse height.

The procedures for preparation of plant chromosome suspensions described here are based on formaldehyde fixation of root tips. The fixation results in lower resolution of flow karyotypes after staining with DNA intercalators such as ethidium bromide or propidium iodide. Always use DAPI or similar dyes (e.g., Hoechst 33258) to stain chromosomes isolated according to this protocol for high-resolution univariate flow karyotyping.

The resolution of individual chromosomes may be improved after bivariate analysis of chromosome DNA content and the amount of particular repetitive DNA sequences labeled by fluorescence *in situ* hybridization in suspension (FISHIS, Giorgi et al. 2013). To date, successful use of FISHIS was reported in wheat and its wild relatives (Akpinar et al. 2015a; Molnár et al. 2016). Molnár et al. (2016) showed that chromosome discrimination could be further improved by using several different probes for FISHIS simultaneously. Unfortunately, it seems that FISHIS is not universally applicable and fails in some plant species (data not published). In our hands, the most successful probes are those based on microsatellite sequences, while rDNA or centromeric sequences do not provide satisfying results. The suitability of FISHIS needs to be tested for each probe and plant species.

5.16.8.4 Chromosome Sorting

Sorting of pure chromosome fractions requires that the sorter be well aligned. In addition to the resolution of fluorescence and forward scatter (see above), this means adjustment of the sort module. Always check its performance prior to actual chromosome sorting. Adjust the drop drive phase to obtain single side streams without fanning, and make sure that the sorted stream is positioned so that the deflected drops arrive at the collecting tube or on a slide. Check the adjustment of the drop delay so that the number of sorted particles indicated by the machine is equal to that actually sorted. Carefully follow the stability of the break-off point and immediately stop the sorting if a change is observed. If this happens, the sort module must be readjusted. Note that a common reason for break-off point instability is a dirty nozzle. Many modern instruments are equipped with automated sort control systems which monitor and adjust the correct drop delay and in case of emergency (e.g., clogged nozzle) stop the sorting. Nevertheless, operator should be always ready for rapid reaction to avoid unnecessary contamination of the sorted fractions.

Although the purity of sorted chromosome fractions is primarily determined by the resolution of flow karyotypes, it is also affected by the presence of debris particles, chromatids, and chromosome clumps in the chromosome suspension. For instance, chromatids and fragments of larger chromosomes may be indiscernible from small chromosomes or chromosome arms. It may be difficult to discriminate doublets of short chromosomes or other chromosome aggregates from large chromosomes. Although the analysis of fluorescence pulse width may be helpful (see basic protocol 3 and alternate protocol 4), it is preferable to minimize the frequency of these particles.

Always check the purity of sorted chromosome fractions. Because the degree of condensation varies greatly among sorted chromosomes and hence also chromosome morphology may vary, simple fluorescence staining is not sufficient to determine the purity of a sorted chromosome fraction. This is best done after FISH with sorted chromosomes using either chromosome-specific DNA sequences or sequences that show a chromosome-specific labeling pattern (see support protocol 3). When sorting, the instrument should be triggered on the fluorescence pulse height signal and the threshold set to a minimum so that all DNA-containing particles are detected. This is important for recognition of particles that should not be sorted, and for avoiding contamination of the sorted fraction. Even if the sample line is equilibrated with dye, a slight change in chromosome peak position may be observed, especially during long-term sorting. Monitor peak positions during sorting and make adjustments to the sort window as needed.

5.16.8.5 Estimation of Purity of Sorted Fractions using FISH

Characterization of individual chromosome types is done using FISH with fluorescently labeled DNA probes. The fluorescent labeling can be direct (the fluorescent molecule is directly bound to the probe) or indirect (the probe is labeled using biotin or a hapten and then visualized using fluorescently labeled avidin/streptavidin, or antibodies against the hapten). Although the direct labeling makes FISH protocol faster and more convenient, it has also its own weaknesses. When comparing directly and indirectly labeled probes, the signal of the direct probe is significantly weaker. In many applications this is not an essential problem, nevertheless, chromosome identification is not a trivial task and presence or absence of one weak signal may change the chromosome identity. Thus, attention must be paid to selecting appropriate probes.

When performing FISH for purity estimation, the slides should be bubble-free in every step. Bubbles prevent the antibody or probe to hybridize to the target, which can cause false interpretations. Similarly, the probe used as marker should be robust and reliable. Due to the lower hybridization efficiency, short unique sequences are not suitable candidates for this application. In our laboratory, we use mainly repetitive DNA labeled using PCR or nick translation.

Note: In case of purity estimation of FISHIS-labeled chromosomes, it is not necessary but advisable to redo FISH labeling with more probes for better identification of chromosomes.

5.16.8.6   Multiple Displacement Amplification of Chromosomal DNA

The optimal extent of proteinase treatment, mainly proteinase K concentration, depends on chromatin structure and fixation conditions, and must be optimized for each plant species. The presented protocol has been optimized for cereals. Based on our experience, fescue, chickpea or corn require increased proteinase K concentration (as much as five times).

The most critical step is DNA purification on columns. Do not allow the whole sample volume to run through the column; the column membrane should not dry during the centrifugation. On the other hand, the final volume of the purified sample should not exceed 20 μL. Too extensive reduction of sample volume (more than seven times) results in decreased amplification yield, or even complete inhibition of the amplification.

5.16.8.7   Single Chromosome Sorting and Amplification

When sorting and amplifying single chromosomes, special care must be taken to avoid contamination by alien DNA. For the sterility reasons (easier to sterilize), we use a ready-to-use solution of proteinase K from Sigma-Aldrich (Cat. no. P4850). It is also critical to decontaminate the flow cytometer before use. Modern flow sorters are equipped with automated procedures (e.g., "prepare for aseptic sort" in the case of FACSAria). Always work in a presterilized biosafety cabinet. Treat every piece of lab ware and reagents (except enzymes) with UV light. Use pipette tips with filters, only. It is advisable not to submerge the pipette tip into the reaction, as DNA might stick to it, but rather pipette on the tube wall and spin down immediately. Although digested chromosomes can be stored at −20°C after the proteinase treatment, it is best to process them within few weeks, as the DNA may degrade. The amplification time can be shortened to 3 h for large chromosomes (e.g., some cereal or legume chromosomes) or prolonged up to 5 h in the case of small compact chromosomes (e.g., corn B chromosomes). Initial proteinase K concentration (10:1) may be raised to 10:3 when amplifying highly heterochromatic chromosomes; the amplification time has also to be raised in these cases.

5.16.8.8   Preparation of HMW DNA

The amount of collection medium depends on the volume of the droplet containing the sorted chromosome and the number of chromosomes sorted. For example, our FACSAria equipped with a 70-μm nozzle and run at 70 psi pressure produces droplets of 1.1 nl volume. Thus, the appropriate amount of collection media for 700,000 chromosomes is 770 μL.

If the required number of chromosomes cannot be reached in one sorting day, store the tube with flow-sorted chromosomes at 4°C and complete the sorting on the following day. Proceed to DNA purification immediately. Avoid freezing.

For plant species rich in polyphenolic compounds, increase the mercaptoethanol concentration in the 1.5× IB three times. In addition, add mercaptoethanol at the same concentration in the lysis buffer C.

5.16.8.9   Proteomic Analyses

To avoid degradation of chromosomal proteins, PMSF is added to the collection media before the sorting. The collection tubes should be kept on ice during sorting. If the required number of chromosomes cannot be reached in one sorting day, store the tube at −20°C until the next sorting experiment.

5.17   ANTICIPATED RESULTS

5.17.1   Preparation of Chromosome Suspensions

Depending on the species, the procedure for metaphase accumulation should result in metaphase frequencies ranging between 40% and 60%. This level of cell cycle synchrony is essential for preparation of good chromosome suspensions. Besides the degree of cell synchrony, the chromosome yield depends on the strength of formaldehyde fixation and the procedure for chromosome release itself. Under optimal conditions, $>10^6$ chromosomes can be isolated from 30 root tips of field bean, garden pea, or hexaploid wheat. Approximately 5×10^5 chromosomes can be isolated after homogenization of 50 root tips of rye and barley.

5.17.2   Flow Karyotyping and Chromosome Sorting

Figure 5.3 presents typical univariate flow karyotypes obtained after the analysis of DAPI-stained chromosomes isolated from hexaploid bread wheat ($2n = 6x = 42$). Root-tip meristems were synchronized according to basic protocol 1, chromosomes were isolated according to basic protocol 2, and sorted according to basic protocol 3. Figure 5.3a shows a flow karyotype obtained in a line with a standard karyotype. Note that only chromosome 3B can be fully resolved, while the remaining chromosomes form three composite peaks. Figure 5.3b demonstrates the flow karyotype of the cultivar Cezanne with translocation chromosome 5BL/7BL. Figure 5.3 shows an

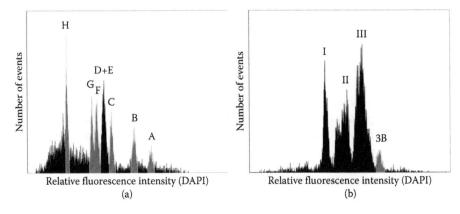

FIGURE 5.3 Flow karyotyping in chickpea and bread wheat. The fluorescence intensity histograms (flow karyotypes) were obtained from DAPI-stained suspensions of mitotic chromosomes. (a) Chickpea cv. Frontier ($2n = 2x = 16$) forms seven peaks, six of which each represent a single chromosome (A–C and F–H). The seventh peak harbors both chromosomes D and E. (b) In the wheat cv. Chinese Spring ($2n = 6x = 42$) flow karyotype, only chromosome 3B forms a discrete peak. The remaining 20 chromosomes are dispersed into the composite peaks I–III. (From Doležel et al. *Biotechnology Advances* 32, 122–136, 2017.)

example of a deletion line with telocentric chromosome 7DS (short arm of chromosome 7D). It also depicts an alien addition line of wheat carrying barley chromosome 7HS. Note that the flow karyotypes also contain minor peaks representing chromosomal debris, chromatids, and chromosome clumps.

The results of bivariate flow karyotyping in tetraploid ($2n = 4x = 28$) durum wheat (cv. Creso) after FISHIS are demonstrated in Figure 5.3a. Chromosome suspensions were prepared according to basic protocols 1 and 2, and then labeled using FISHIS, stained with DAPI, and analyzed according to alternate protocol 4. The combination of two parameters, DNA content as measured by intensity of DAPI fluorescence and amount of GAA microsatellites (FITC signal), allows for unambiguous discrimination of 12 out of 14 chromosomes. The durum wheat genome is composed of two subgenomes, A and B. Note that chromosomes of the B subgenome have a higher content of GAA sequences.

In the flow karyotype of chickpea, six of the eight chromosomes can be separated, while the other two form a single peak (Figure 5.3a). In contrast, the bread wheat (n = 21) flow karyotype comprises only one single chromosome peak (chromosome 3B), with the other 20 chromosomes forming three composite peaks (Figure 5.3b).

5.17.3 Determining the Purity of Sorted Chromosome Fractions Using FISH

The purity in flow-sorted fractions depends on sample quality (presence of chromatids and chromosome fragments and clumps) and the ability to discriminate the chromosome of interest from remaining chromosomes in the karyotype. The purities of sorted chromosome fractions with well-resolved peaks on flow karyotypes can approach 100%. For example, when sorting chromosomes of bread wheat, Vrána et al. (2000) and Kubaláková et al. (2002) achieved purities of 95% and 97% for chromosomes 3B and 5BL/7BL, respectively. On the other hand, purities of chromosomes sorted from composite peaks may be as low as a few percent, as shown for chromosome 7B of bread wheat (Vrána et al. 2015). Figure 5.3b shows an example of a purity check of flow-sorted chromosomes of durum wheat cv. Creso identified after FISH with a probe for GAA microsatellites and *Afa* repeat family.

Population of chromosome 1B was sorted from the region highlighted in Figure 5.3a. Out of the five chromosomes in the picture, four were identified as chromosome 1B with one copy of chromosome 4B as a contamination. At least 100 chromosomes should be evaluated for a reliable purity check.

5.17.4 DNA AMPLIFICATION

Purification of DNA from flow-sorted chromosomes typically results in yields of 50% of the amount of DNA in flow-sorted chromosomes, that is, 20–30 ng if the equivalent of 50 ng DNA was sorted. This amount of purified

DNA provides the template for three amplification reactions (7 ng–10 ng/reaction). Pooling of three amplification products is recommended to reduce amplification bias. The majority of amplification products is between 5–20 kb, which is suitable for short-read NGS technologies. The multiple displacement amplification can lead to chimera formation, which may potentially cause problems in mate-pair sequencing and long-read sequencing technologies so the protocol is not recommend for these applications.

5.17.5 PREPARATION OF HMW DNA

The protocol provides DNA of megabase size, which is well accessible to restriction endonucleases and nicking enzymes. The amount of DNA released from agarose miniplugs varies from 30% to 60% of the flow-sorted amount and depends largely on the execution of the supernatant removal step. The quality of the resulting HMW DNA can be tested using pulsed field gel electrophoresis (PFGE). Depending on PFGE conditions, the DNA is mostly observed in the well and compression zone of the gel.

5.17.6 TIME CONSIDERATIONS

The whole procedure, from the preparation of chromosome suspensions to their analysis and chromosome sorting, may take up to one week. Thus, the experiment needs careful planning. Although the protocols described here are not complicated, they require expertise in several different areas.

5.17.7 PREPARATION OF CHROMOSOME SUSPENSIONS

The quality of chromosome suspensions is critical for successful chromosome sorting, and it may take several weeks to master the preparation method. The chromosome yield depends critically on the frequency of metaphase cells in the root tip. In some cases, it may be necessary to optimize the protocol for cell cycle synchronization and metaphase accumulation. Generally, the preparation of plant material from seed germination to accumulation of root tips in metaphase takes 5–6 days. It is convenient to germinate the seeds over a weekend. The processing of root tips for chromosome isolation is not time-consuming and takes ~1 h including formaldehyde fixation. However, it must be performed immediately after metaphase accumulation.

5.17.8   FISHIS

FISHIS is a straightforward and rapid procedure. Denaturation of chromosomes takes 20 min; incubation with a probe takes 1 h. Altogether, with hands-on time the complete protocol takes about 90 min.

5.17.9   FLOW ANALYSIS AND SORTING

The operation of the flow cytometer for flow karyotyping and chromosome sorting requires a trained specialist. The analysis must be performed at the highest possible resolution. This often requires time-consuming alignment of the instrument optics and fluidics. If the flow cytometer is well aligned and its fluidics are stable, chromosome sorting can be routinely performed at 20–40 chromosomes/s. Naturally, the actual sort rate depends on the quality of the suspension, the frequency of the chromosome in the suspension, and the sort mode. Nevertheless, sorting of small amounts of chromosomes for PCR or microscopic evaluation is convenient and takes only a few minutes. Sorting appropriate numbers of chromosomes needed for DNA amplification requires several hours depending on molecular size (DNA content) of chromosomes. Here problems with flow stability, including occasional clogging of the nozzle, may result in additional delays. The most demanding applications regarding sorting time are preparation of HMW DNA and proteomic analyses. The chromosomes are sorted in batches and it can take several working days or even weeks to obtain sufficient amounts of chromosomal DNA, or proteins.

5.17.10   ESTIMATING THE PURITY OF SORTED CHROMOSOMES

Before the procedure, the sorted chromosomes must be air dried on a microscopic slide. Although this may take only ~1 h, it may be convenient to dry the slides overnight. The FISH procedure takes 2 days. On the first day, denaturation takes place, followed by overnight hybridization with probes (either direct or indirect). On the second day, the procedure takes ~1.5 h and 3 h in case of direct and indirect labeling, respectively. Microscopic evaluation of one slide (analyzing ~100 chromosomes) takes ~30 min.

5.17.11   SINGLE CHROMOSOME DNA AMPLIFICATION

The whole procedure can be completed in 2 days. Immediately after sorting the chromosomes are treated overnight (at least 18 h) at 50°C to digest proteins.

Inactivation of proteinase and at least one freeze/thaw cycle takes about an hour. Preparation of a sterile working environment takes half an hour, hands-on time to set up the reactions depends on number of samples to be processed, usually between half an hour and one hour. The reaction itself takes four-and-a-half hours. Quality control of the resulting amplicons takes about four hours (purification, electrophoresis, concentration measurements).

5.17.12   PREPARATION OF HMW DNA

The whole procedure takes 2 days. Several batches of flow-sorted chromosomes can be processed together. The procedure of plug preparation takes ~1 h, DNA purification ~44 h (30 min hands-on work).

5.17.13   PROTEOMIC ANALYSES

The whole procedure can be completed in 5 days. Several batches of flow-sorted chromosomes can be processed together. On the first day the extraction of chromosomal proteins takes place, on the second day the protein separation and detection is performed. Proteins are digested on the third day, desalted on the fourth, and the MS analyses are run on the fifth day.

REFERENCES

Akpinar, B. A., S. J. Lucas, J. Vrána, J. Doležel, and H. Budak, 2015b. Sequencing chromosome 5D of *Aegilops tauschii* and comparison with its allopolyploid descendant bread wheat (*Triticum aestivum*). *Plant Biotech. J.* 13: 740–752. doi:10.1111/pbi.12302.

Akpinar, B. A., M. Yuce, S. Lucas, J. Vrána, V. Burešová, J. Doležel, and H. Budak, 2015a. Molecular organization and comparative analysis of chromosome 5B of the wild wheat ancestor *Triticum dicoccoides*. *Sci. Rep.* 5: 10763. doi:10.1038/srep10763.

Arumuganathan, K., G. B. Martin, H. Telenius, S. D. Tanksley, and E. D. Earle, 1994. Chromosome 2-specific DNA clones from flow-sorted chromosomes of tomato. *Mol. Gen. Genet.* 242: 551–558. doi: 10.1007/BF00285278.

Arumuganathan, K., J. P. Slattery, S. D., Tanksley, and E. D. Earle, 1991. Preparation and flow cytometric analysis of metaphase chromosomes of tomato. *Theor. Appl. Genet.* 82: 101–111. doi:10.1007/BF00231283.

Berkman, P.J., A. Skarshewski, S. Manoli, M. T. Lorenc, J. Stiller, L. Smits, K. Lai, E. Campbell, M. Kubaláková, H. Šimková, J. Batley, J. Doležel, P. Hernandez, and D. Edwards, 2012. Sequencing wheat chromosome arm 7BS delimits the 7BS/4AL translocation and reveals homoeologous gene conservation. *Theor. Appl. Genet.* 124: 423–432. doi: 10.1007/s00122-011-1717-2.

Binarová, P., B. Hause, J. Doležel, and P. Dráber, 1998. Association of γ-tubulin with kinetochore/centromeric region of plant chromosomes. *Plant J.* 14: 751–757. doi: 10.1046/j.1365-313x.1998.00166.x.

Boschman, G. A., E. M. M. Manders, W. Rens, R. Slater, and J. A. Aten, 1992. Semi-automated detection of aberrant chromosomes in bivariate flow karyotypes. *Cytometry* 13: 469–477. doi:10.1002/cyto.990130504.

Cápal, P., N. Blavet, J. Vrána, M. Kubaláková, and J. Doležel, 2015. Multiple displacement amplification of the DNA from single flow-sorted plant chromosome. *Plant J.* 84: 838–844. doi: 10.1111/tpj.13035.

Cápal, P., T. R. Endo, J. Vrána, M. Kubaláková, M. Karafiátová, E. Komínková, I. Mora-Ramírez, W. Weschke, and J. Doležel, 2016. The utility of flow sorting to identify chromosomes carrying a single copy transgene in wheat. *Plant Meth.* 12: 24. doi: 10.1186/s13007-016-0124-8.

Carter, N. P. 1994. Cytogenetic analysis by chromosome painting. *Cytometry* 18: 2–10. doi: 10.1002/cyto.990180103.

Chen, W., V. Kalscheuer, A. Tzschach, C. Menzel, R. Ullmann, M. H. Schulz, F. Erdogan, et al. 2008. Mapping translocation breakpoints by next generation sequencing. *Genome Res.* 18: 1143–1149. doi: 10.1101/gr.076166.108.

Chen, W., R. Ullmann, C. Langnick, C. Menzel, Z. Wotschofsky, H. Hu, A. Döring, et al. 2010. Breakpoint analysis of balanced chromosome rearrangements by next generation paired-end sequencing. *Eur. J. Hum. Genet.* 18: 539–543. doi: 10.1038/ejhg.2009.211.

Conia, J., C. Bergounioux, S. Brown, C. Perennes, and P. Gadal, 1988. Caryotype en flux biparametrique de *Petunia hybrida*. Tri du chromosome numero I. *C.R. Acad. Sci.* 307: 609–615.

Conia, J., C. Bergounioux, C. Perennes, P. Muller, S. Brown, and P. Gadal, 1987. Flow cytometric analysis and sorting of plant chromosomes from *Petunia hybrida* protoplasts. *Cytometry* 8: 500–508. doi: 10.1002/cyto.990080511.

Conia, J., P. Muller, S. Brown, C. Bergounioux, and P. Gadal, 1989. Monoparametric models of flow cytometric karyotypes with spread sheet software. *Theor. Appl. Genet.* 77: 295–303. doi:10.1007/BF00266200.

Cooke, A., J. L. Tolmie, J. M. Colgan, C. M. Greig, and J. M. Connor, 1989. Detection of an unbalanced translocation (4; 14) in a mildly retarded father and son by flow cytometry. *Hum. Genet.* 83: 83–87. doi: 10.1007/BF00274155.

De Laat, A. M. M. and J. Blaas, 1984. Flowcytometric characterization and sorting of plant chromosomes. *Theor. Appl. Genet.* 67: 463–467. doi: 10.1007/BF00263414.

Doležel, J., P. Binarová, and S. Lucretti, 1989. Analysis of nuclear DNA content in plant cells by flow cytometry. *Biol. Plant.* 31: 113–120. doi:10.1007/BF02907241.

Doležel, J., J. Číhalíková, and S. Lucretti, 1992. A high-yield procedure for isolation of metaphase chromosomes from root tips of *Vicia faba* L. *Planta* 188: 93–98. doi: 10.1007/BF00198944.

Doležel, J., J. Číhalíková, J. Weiserová, and S. Lucretti, 1999. Cell cycle synchronization in plant root meristems. *Methods Cell Sci.* 21: 95–107. doi: 10.1023/A:1009876621187.

Doležel, J. and S. Lucretti, 1995. High-resolution flow karyotyping and chromosome sorting in *Vicia faba* lines with standard and reconstructed karyotypes. *Theor. Appl. Genet.* 90: 797–802. doi: 10.1007/BF00222014.

Doležel, J., S. Lucretti, and I. Schubert, 1994. Plant chromosome analysis and sorting by flow cytometry. *Crit. Rev. Plant Sci.* 13: 275–309. doi: 10.1080/07352689409701917.

Doležel, J., M. A. Lysák, M. Kubaláková, H. Šimková, J. Macas, and S. Lucretti, 2001. Sorting of plant chromosomes. In Methods in Cell Biology, Vol. 64. 3rd Edition, Part B. (Z. Darzynkiewicz, H.A. Crissman, and J.P. Robinson, (Eds.). pp. 3–31. Academic Press, New York.

Doležel, J., J. Vrána, P. Cápal, M. Kubaláková, V. Burešová, and H. Šimková, 2014. Advances in plant chromosome genomics. *Biotechnol. Adv.* 32: 122–136. doi: 10.1016/j.biotechadv.2013.12.011.

Doležel, J., J. Vrána, J. Šafář, J. Bartoš, M. Kubaláková, and H. Šimková, 2012. Chromosomes in the flow to simplify genome analysis. *Funct. Integr. Genomics* 12: 397–416. doi: 10.1007/s10142-012-0293-0.

Endo, T. R., M. Kubaláková, J. Vrána, and J. Doležel, 2014. Hyperexpansion of wheat chromosomes sorted by flow cytometry. *Genes Genet. Syst.* 89: 181–185. doi: 10.1266/ggs.89.181.

Ferguson-Smith, M. A., and V. Trifonov, 2007. Mammalian karyotype evolution. *Nat. Rev. Genet.* 8: 950–962. doi: 0.1038/nrg2199.

Fiegler, H., S. M. Gribble, D.C. Burford, P. Carr, E. Prigmore, K.M. Porter, S. Clegg, et al. 2003. Array painting: A method for the rapid analysis of aberrant chromosomes using DNA microarrays. *J. Med. Genet.* 40: 664–670. doi: 10.1136/jmg.40.9.664.

Fluch, S., D. Kopecky, K. Burg, H. Šimková, S. Taudien, A. Petzold, M. Kubaláková, et al. 2012. Sequence composition and gene content of the short arm of rye (*Secale cereale*) chromosome 1. *PLoS One* 7: e30784. doi: 10.1371/journal.pone.0030784.

Fuchs, J., S. Joos, P. Lichter, and I. Schubert, 1994. Localization of vicilin genes on field bean chromosome II by fluorescent in situ hybridization. *J. Hered.* 85: 487–488.

Gamborg, O. L. and L. R. Wetter, 1975. *Plant Tissue Culture Methods.* Natl. Res. Council of Canada, Saskatoon, Saskatchewan, Canada.

Giorgi, D., A. Farina, V. Grosso, A. Gennaro, C. Ceoloni, and S. Lucretti, 2013. FISHIS: Fluorescence *in situ* hybridization in suspension and chromosome flow sorting made easy. *PLoS One* 8: e57994. doi: 10.1371/journal.pone. 0057994.

Grosso, V., A. Farina, A. Gennaro, D. Giorgi, and S. Lucretti, 2012. Flow sorting and molecular cytogenetic identification of individual chromosomes of *Dasypyrum villosum* L. (*H. villosa*) by single DNA probe. *PLoS One* 7: e50151. doi: 10.1371/journal. pone.0050151.

Gualberti, G., J. Doležel, J. Macas, and S. Lucretti, 1996. Preparation of pea (*Pisum sativum* L) chromosome and nucleus suspensions from single root tips. *Theor. Appl. Genet.* 92: 744–751. doi: 10.1007/BF00226097.

Guo, D. W., Y. Z. Ma, L. C. Li, and Y. F. Chen, 2006. Flow sorting of wheat chromosome arms from the ditelosomic line 7BL. *Plant Mol. Biol. Rep. 24*: 23–31. doi: 10.1007/ BF02914043.

Guo, D. W., D. H. Min, Z. S. Xu, M. Chen, L. C. Li, M. Ashraf, A. Ghafoor, and Y. Z. Ma, 2013. Flow karyotyping of wheat addition line 'T240' with a *Haynaldia villosa* 6VS telosome. *Plant Mol. Biol. Rep.* 31: 289–295. doi: 10.1007/s11105-012-0492-9.

Hernandez, P., M. Martis, G. Dorado, M. Pfeifer, S. Gálvez, S. Schaaf, N. Jouve, et al. 2012. Next-generation sequencing and syntenic integration of flow-sorted arms of wheat chromosome 4 A exposes the chromosome structure and gene content. *Plant J.* 69: 377–386. doi:10.1111/j.1365-313X.2011.04808.x.

Janda, J., J. Šafář, M. Kubaláková, J. Bartoš, P. Kovářová, P. Suchánková, S. Pateyron, et al. 2006. Advanced resources for plant genomics: BAC library specific for the short arm of wheat chromosome 1B. *Plant J.* 47: 977–986. doi: 10.1111/j.1365-313X.2006.02840.x.

Kato, A. 1999. Air drying method using nitrous oxide for chromosome counting in maize. *Biotech. Histochem.* 74: 160–166. doi: 10.3109/10520299909047968.

Kejnovský, E., J. Vrána, S. Matsunaga, P. Souček, J. Široký, J. Doležel, and B. Vyskot, 2001. Localization of male-specifically expressed MROS genes on *Silene latifolia* by PCR on flow sorted sex chromosomes and autosomes. *Genetics* 158: 1269–1277.

Kopecký, D., M. Martis, J. Čížková, E. Hřibová, J. Vrána, J. Bartoš, J. Kopecká, et al. 2013. Flow sorting and sequencing meadow fescue chromosome 4 F. *Plant Physiol.* 163: 1323–1337. doi: 10.1104/pp.113.224105.

Korstanje, R., G. F. Gillissen, M. G. den Bieman, S. A. Versteeg, B. van Oost, R. R. Fox, H. A. van Lith, and van L. F. Zutphen, 2001. Mapping of rabbit chromosome 1 markers generated from a microsatellite-enriched chromosome specific library. *Anim. Genet.* 32: 308–312. doi: 10.1046/j.1365-2052.2001.00783.x.

Kubaláková, M., P. Kovářová, P. Suchánková, J. Číhalíková, J. Bartoš, S. Lucretti, N. Watanabe, S. F. Kianian, and J. Doležel, 2005. Chromosome sorting in tetraploidwheat and its potential for genome analysis. *Genetics* 170: 823–829. doi:10.1534/genetics.104.039180.

Kubaláková, M., M. A. Lysák, J. Vrána, H. Šimková, J. Číhalíková, and J. Doležel, 2000. Rapid identification and determination of purity of flow-sorted plant chromosomes using C-PRINS. *Cytometry* 41: 102–108. doi: 10.1002/ 1097-0320(20001001)41:2%3c102::AIDCYTO4% 3e3.0.CO;2-H.

Kubaláková, M., J. Macas, and J. Doležel, 1997. Mapping of repeated DNA sequences in plant chromosomes by PRINS and CPRINS. *Theor. Appl. Genet.* 94: 758–763. doi:10.1007/s001220050475.

Kubaláková, M., M. Valárik, J. Bartoš, J. Vrána, J. Číhalíková, J. Molnár-Láng, and J. Doležel, 2003. Analysis and sorting of rye (*Secale cereal* L) chromosomes using flow cytometry. *Genome* 46: 893–905. doi: 10.1139/g03-054.

Kubaláková, M., J. Vrána, J. Číhalíková, H. Šimková, and J. Doležel, 2002. Flow karyotyping and chromosome sorting in bread wheat (*Triticum aestivum* L). *Theor. Appl. Genet.* 104: 1362–1372. doi: 10.1007/s00122-002-0888-2.

Lebo, R. V. 1982. Chromosome sorting and DNA sequence localization: A review. *Cytometry* 3: 145–154. doi: 10.1002/cyto.990030302.

Lee, J. H., K. Arumuganathan, S. M. Kaeppler, H. F. Kaeppler, and C. M. Papa, 1996. Cell synchronization and isolation of metaphase chromosomes from maize (*Zea mays* L). root tips for flow cytometric analysis and sorting. *Genome* 39: 697–703. doi: 10.1139/g96-088.

Lee, J. H., K. Arumuganathan, Y. Yen, S. Kaeppler, H. Kaeppler, and P. S. Baezinger, 1997. Root tip cell cycle synchronization and metaphase chromosome isolation suitable for flow sorting in common wheat (*Triticum aestivum* L). *Genome* 40: 633–638. doi: 10.1139/g97- 083.

Lucretti, S., J. Doležel, I. Schubert, and J. Fuchs, 1993. Flow karyotyping and sorting of *Vicia faba* chromosomes. *Theor. Appl. Genet.* 85: 665–672. doi: 10.1007/BF00225003.

Lucretti, S. and J. Doležel, 1997. Bivariate flow karyotyping in broad bean (*Vicia faba*). *Cytometry* 28: 236–242. doi: 10.1002/(SICI) 1097-0320(19970701)28:3%3c236::AID CYTO8% 3e3.0.CO;2-B.

Lysák, M. A., J. Číhalíková, M. Kubaláková, H. Šimková, G. Kűnzel, and Doležel, J. 1999. Flow karyotyping and sorting of mitotic chromosomes of barley (*Hordeum vulgare* L). *Chrom. Res.* 7: 431–444. doi: 10.1023/ A:1009293628638.

Macas, J., J. Doležel, G. Gualberti, U. Pich, I. Schubert, and S. Lucretti, 1995. Primer-induced labeling of pea and field bean chromosomes *in situ* and in suspension. *BioTechniques* 19: 402–408.

Macas, J., J. Doležel, S. Lucretti, U. Pich, A. Meister, J. Fuchs, and I. Schubert, 1993. Localization of seed storage protein genes on flow-sorted field bean chromosomes. *Chrom. Res.* 1: 107–115. doi: 10.1007/BF00710033.

Macas, J., G. Gualberti, M. Nouzová, P. Samec, S. Lucretti, and J. Doležel, 1996. Construction of chromosome-specific DNA libraries covering the whole genome of field bean (*Vicia faba* L). *Chrom. Res.* 4: 531–539. doi: 10.1007/BF02261781.

Martis, M. M., S. Klemme, A. M. Banaei-Moghaddam, F. R. Blattner, J. Macas, T. Schmutzer, U. Scholz, et al. 2012. Selfish supernumerary chromosome reveals its origin as a mosaic of host genome and organellar sequences. *Proc.Natl. Acad. Sci. U.S.A.* 109: 13343–13346. doi: 10.1073/pnas.1204237109.

Martis, M. M., R. Zhou, G. Haseneyer, T. Schmutzer, J. Vrána, M. Kubaláková, S. König, et al. 2013. Reticulate evolution of the rye genome. *Plant Cell* 25: 3685–3698. doi: 10.1105/tpc.113.114553.

Mayer, K. F. X., M. Martis, P. E. Hedley, H. Šimková, H. Liu, J. A. Morris, B. Steuernagel, et al. 2011. Unlocking the barley genome by chromosomal and comparative genomics. *Plant Cell* 23: 1249–1263. doi: 10.1105/tpc.110.082537.

Mayer, K. F. X., J. Rogers, J. Doležel, C. Pozniak, K. Eversole, C. Feuillet, B. Gill, et al. 2014. A chromosome-based draft sequence of the hexaploid bread wheat (*Triticum aestivum*) genome. *Science* 345: 1251788. doi:10.1126/science.1251788.

Mayer, K. F. X., S. Taudien, M. Martis, H. Šimková, P. Suchánková, H. Gundlach, T. Wicker, et al. 2009. Gene content and virtual gene order of barley chromosome 1 H. *Plant Physiol.* 151: 496–505. doi: 10.1104/pp.109 .142612.

Molnár, I., M. Kubaláková, H. Šimková, A. Cseh, M. Molnár-Láng, and J. Doležel, 2011. Chromosome isolation by flow sorting in *Aegilops umbellulata* and *Ae. comosa* and their allotetraploid hybrids *Ae. biuncialis and Ae. geniculata.PLoS One* 6: e27708. doi: 10.1371/journal.pone.0027708.

Molnár, I., M. Kubaláková, H. Šimková, A. Farkas, A. Cseh, M. Megyeri, J. Vrána, M. Molnár-Láng, and J. Doležel, 2014. Flow cytometric chromosome sorting from diploid progenitors of bread wheat, *T. urartu, Ae. speltoides and Ae. tauschii. Theor. Appl. Genet.* 127: 1091–1104. doi: 10.1007/s00122-014-2282-2.

Molnár, I., J. Vrána, V. Burešová, P. Cápal, A. Farkas, E. Darkó, A. Cseh, M. Kubaláková, M. Molnár-Láng, and J. Doležel, 2016. Dissecting the U, M, S and C genomes of wild relatives of bread wheat (*Aegilops* spp.) into chromosomes and exploring their macrosyntheny with wheat. *Plant J.* 88(3): 452–467.

Molnár, I., J. Vrána, A. Farkas, M. Kubaláková, A. Cseh, M. Molnár-Láng, and J. Doležel, 2015. Flow sorting of C-genome chromosomes from wild relatives of wheat *Aegilops markgrafii.Ae. triuncialis* and *Ae. cylindrica, and their molecular organization. Ann. Bot.* 116: 189–200. doi: 10.1093/aob/mcv073.

Neumann, P., M. Lys´ak, J. Dole˘zel, and J. Macas, 1998. Isolation of chromosomes from *Pisum sativum* L. hairy root cultures and their analysis by flow cytometry. *Plant Sci.* 137: 205–215. doi:10.1016/S0168-9452(98)00141-1.

Neumann, P., D. Požárková, J. Vrána, J. Doležel, and J. Macas, 2002. Chromosome sorting and PCR-based physical mapping in pea (*Pisum sativum* L). *Chrom. Res.* 10 63–71. doi: 10.1023/A:1014274328269.

Nie, X., B. Li, L. Wang, P. Liu, S. S. Biradar, T. Li, J. Doležel, D. Edwards, M. Luo, and S. Weining, 2012. Development of chromosome-armspecific microsatellite markers in *Triticum aestivum* (Poaceae) using NGS technology. *Am. J. Bot.* e369–e371. doi: 10.3732/ajb.1200077.

Otto, F. J. 1988. Assessment of persisting chromosome aberrations by flow karyotyping of cloned Chinese hamster cells. *Z. Naturforsch.* 43c: 948–954.

Petrovská, B., H. Jeřábková, I. Chamrád, J. Vrána, R. Lenobel, J. Uřinovská, M. Sebela, and J. Doležel, 2014. Proteomic analysis of barley cell nuclei purified by flow sorting. *Cytogenet. Genome Res.* 143: 78–86. doi:10.1159/000365311.

Pich, U., A. Meister, J. Macas, J. Doležel, S. Lucretti, and I. Schubert, 1995. Primed *in situ* labelling facilitates flow sorting of similar sized chromosomes. *Plant J.* 7: 1039–1044. doi: 10.1046/j.1365-313X.1995.07061039.x.

Ruperao, P., C. K. Chan, S. Azam, M. Karafiátová, S. Hayashi, J. Čížková, R.K. Saxena, et al.2014. A chromosomal genomics approach to assess and validate the *desi* and *kabuli* draft chickpea genome ssemblies. *Plant Biotechnol. J.* 12: 778–786. doi: 10.1111/pbi.12182.

Šafář, J., J. Bartoš, J. Janda, A. Bellec, M. Kubaláková, M. Valárik, S. Pateyron, et al. 2004. Dissecting large and complex genomes: Flow sorting and BAC cloning of individual chromosomes from bread wheat. *Plant J.* 39: 960–968. doi:10.1111/j.1365-313X.2004.02179.x.

Šafář, J., H. Šimková, M. Kubaláková, J. Číhalíková, P. Suchánková, J. Bartoš, and J. Doležel, 2010. Development of chromosome-specific BAC resources for genomics of bread wheat. *Cytogenet. Genome Res.* 129: 211–223. doi: 10.1159/000313072.

Sala, F., B. Parisi, D. Burroni, A. R. Amileni, G. Pedrali-Noy, and S. Spadari, 1980. Specific and reversible inhibition by aphidicolin of the α-like DNA polymerase of plant cells. *FEBS Lett.* 117: 93–98. doi:10.1016/0014-5793(80)80920-3.

Sargan, D. R., B. S. Milne, J. A. Hernandez, P. C. O'Brien, M. A. Ferguson-Smith, T. Hoather, and J. M. Dobson, 2005. Chromosome rearrangements in canine fibrosarcomas. *J. Hered.* 96: 766–773. doi: 10.1093/jhered/esi122.

Sargan, D. R., F. T. Yang, M. Squire, B. S. Milne, P. C. M. O'Brien, and M. A. Ferguson-Smith, 2000. Use of flow-sorted canine chromosomes in the assignment of canine linkage, radiation hybrid, and syntenic groups to chromosomes: Refinement and verification of the comparative chromosome map for dog and human. *Genomics* 69: 182–195. doi: 10.1006/geno.2000.6334.

Schubert, I., J. Doležel, A. Houben, H. Schertan, and G. Wanner, 1993. Refined examination of plant metaphase chromosome structure at different levels made feasible by new isolation methods. *Chromosoma* 102: 96–101. doi:10.1007/BF00356026.

Schwarzacher, T., M. L. Wang, A. R. Leitch, N. Miller, G. Moore, and J. S. Heslop-Harrison, 1997. Flow cytometric analysis of the chromosomes and stability of a wheat cell culture line. *Theor. Appl. Genet.* 94: 91–97. doi:10.1007/s001220050386.

Šimková, H., J. Čáhaláková, J. Vrána, M. A. Lysák, and J. Doležel, 2003. Preparation of HMW DNA from plant nuclei and chromosomes isolated from root tips. *Biol. Plant* 46:369–373. doi:10.1023/A:1024322001786.

Šimková, H., J. Šafář, P. Suchánková, P. Kovářová, J. Bartoš, M. Kubaláková, J. Janda, J. Číhalíková, R. Mago, T. Lelley, and J. Doležl, 2008b. A novel resource for genomics of Triticeae: BAC library specific for the short arm of rye (*Secale cereale* L). chromosome 1R (1RS). *BMC Genomics* 9: 237.

Šimková, H., J. Šafář, M. Kubaláková, P. Suchánková, J. Číhalíková, H. Robert-Quatre, P. Azhaguvel, et al. 2011. BAC Libraries from wheat chromosome 7D: Efficient tool for positional cloning of aphid resistance genes. *J. Biomed. Biotechnol.* 30: 2543.

Šimková, H., J. T. Svensson, P. Condamine, E. Hřibová, P. Suchánková, P. R. Bhat, J. Bartoš, J. Šafář, T. J. Close, and J. Doležel, 2008a. Coupling amplified DNA from flow sorted chromosomes to high-density SNP mapping in barley. *BMC Genomics* 9: 294. doi:10.1186/1471-2164-9-294

Staňková, H., A. R. Hastie, S. Chan, J. Vrána, Z. Tulpová, M. Kubaláková, P. Visendi, et al. 2016. BioNano genome mapping of individual chromosomes supports physical mapping and sequence assembly in complex plant genomes. *Plant Biotechnol. J.* 14: 1523–1531. doi: 10.1111/pbi.12513.

Stap, J., J. A. Aten, D. Lillington, A. Shelling, and B. Young, 2001. Advanced preparative techniques to establish probes for molecular cytogenetics. *Curr. Protoc. Cytom.* 5: 8.6.1–8.6.23. doi: 10.1002/0471142956.cy0806s05.

Suchánková, P., M. Kubaláková, P. Kovářová, J. Bartoš, J. Číhalíková, M. Molnár-Láng, T. R. Endo, and J. Doležel, 2006. Dissection of the nuclear genome of barley by chromosome flow sorting. *Theor. Appl. Genet.* 113: 651–659. doi: 10.1007/s00122-006-0329-8.

The International Wheat Genome Sequencing Consortium. 2014. A chromosome-based draft sequence of the hexaploid wheat (*Triticum aestivum*) genome. *Science* 345: 1251788. doi: 10.1126/science.1251788.

Tiwari, V. K., S. Wang, T. Danilova, D. H. Koo, J. Vrána, M. Kubaláková, E. Hřibová, et al. 2015. Exploring the tertiary gene pool of bread wheat: Sequence assembly and analysis of chromosome 5Mg of *Aegilops geniculata*. *Proc.Natl. Acad. Sci. U.S.A.* 112: 13633–13638. doi: 10.1073/pnas.1512255112.

Tiwari, V. K., S. Wang, S. Sehgal, J. Vrána, B. Friebe, M. Kubaláková, P. Chhuneja, 2014. SNP Discovery for mapping alien introgressions in wheat. *BMC Genomics* 15: 273. doi: 10.1186/1471-2164-15-273.

Überall, I., J. Vräna, J. Bartoš, J. Šmerda, J. Doležel, and L. Havel, 2004. Isolation of chromosomes from *Picea abies* L. and their analysis by flow cytometry. *Biol. Plant.* 48: 199–203. doi: 10.1023/B:BIOP.0000033445.15336.04.

Valärik, M., J. Bartoš, P. Kovářovä, M. Kubaläková, H. de Jong, and J. Doležel, 2004. High resolution FISH on super-stretched flow-sorted plant chromosomes. *Plant J.* 37: 940–950. doi: 10.1111/j.1365-313X.2003.02010.x.

Van Dilla, M. A., and L. L. Deaven, 1990. Construction of gene libraries for each human chromosome. *Cytometry* 11: 208–218. doi : 10.1002/cyto.990110124.

Veltman, I. M., J. A. Veltman, G. Arkesteijn, I. M. Janssen, L. E. Vissers, P. J. de Jong, A. G. van Kessel, and E. F. Schoenmakers, 2003. Chromosomal breakpoint mapping by array-CGH using flow-sorted chromosomes. *Biotechniques* 35: 1066–1070.

Veuskens, J., D. Marie, S. C. Brown, M. Jacobs, and I. Negrutiu, 1995. Flow sorting of the Y sex-chromosome in the dioecious plant *Melandrium album*. *Cytometry* 21: 363–373.

Veuskens, J., D. Marie, S. Hinnisdaels, and S. C. Brown, 1992. Flow cytometry and sorting of plant chromosomes. *In Flow Cytometry and Cell Sorting* A. Radbruch (Ed.). pp. 177–188. Springer-Verlag, Berlin-Heidelberg.

Vláčilová, K., D. Ohri, J. Vrána, J. Číhalíková, M. Kubaláková, G. Kahl, and J. Doležel, 2002. Development of flow cytogenetics and physical genome mapping in chickpea (*Cicer arietinum* L). *Chrom. Res.* 10: 695–706. doi: 10.1023/A:1021584914931.

Vrána, J., M. Kubaláková, J. Číhalíkoví, M. Valírik, and J. Doležel, 2015. Preparation of subgenomic fractions enriched for particular chromosomes in polyploid wheat. *Biol. Plant* 59: 445–455. doi: 10.1007/s10535-015-0522-1.

Vrána, J., M. Kubaláková, H. Šimková, J. Číhalíkoví, M. A. Lysák, and J. Doležel, 2000. Flow-sorting of mitotic chromosomes in common wheat (*Triticum aestivum* L). *Genetics* 156: 2033–2041.

Vrána, J., H. Šimková, M. Kubaláková, J. Číhalíkoví, and J. Doležel, 2012. Flow cytometric chromosome sorting in plants: The next generation. *Methods* 57: 331–337. doi:10.1016/j.ymeth.2012.03.006.

Vrána, J., P. Cápal, H. Šimková, M. Karafiátová, J. Čížková, and J. Doležel, 2016. Flow analysis and sorting of plant chromosomes. *Curr. Protoc. Cytom.* 78: 1–43. doi:10.1002/cpcy.9.

Wang, M. L., A. R. Leitch, T. Schwarzacher, J. S. Heslop-Harrison, and G. Moore, 1992. Construction of a chromosome-enriched HpaII library from flow-sorted wheat chromosomes. *Nucl. Acids Res.* 20: 1897–1901. doi:10.1093/nar/20.8.1897.

Wicker, T., K. F. X. Mayer, H. Gundlach, M. Martis, B. Steuernagel, U. Scholz, H. Šimková, et al. 2011. Frequent gene movement and pseudogene evolution is common to the large and complex genomes of wheat, barley, and their relatives. *Plant Cell* 23: 1706–1718. doi: 10.1105/tpc.111.086629.

Yang, H., X. Chen, and W.H. Wong, 2011. Completely phased genome sequencing through chromosome sorting. *Proc. Natl. Acad Sci. U.S.A.* 108: 12–17. doi:10.1073/pnas.1016725108.

Young, C.W., and S. Hodas, 1964. Hydroxyurea:Inhibitory effect on DNA metabolism. *Science* 146: 1172. doi: 10.1126/science.146.3648.1172.

Zatloukalová, P., E. Hřibová, M. Kubaláková, P. Suchánková, H. Šimková, C. Adoración, G. Kahl, T. Millán, and J. Doležel, 2011. Integration of genetic and physical maps of the chickpea (*Cicer arietinum* L). genome using flow-sorted chromosomes. *Chromosome Res.*: 729–739. doi: 10.1007/s10577-011-9235-2.

6 Pollen Staining

6.1 INTRODUCTION

Pollen grains are generated by microgametophytes of seed plants. These are produced by anthers undergoing meiosis resulting in microspores with three micronuclei (one vegetative and two sperm nuclei) being generated (Figure 6.1). Pollen grains are powdery and their numbers depend upon the nature of plants; if cross-pollinated, anthers extrude (like in maize) and produce large numbers of pollen grains (Figure 6.2; http://www.science20.com/genetic_maize/those_ naughty_plants). These pollens are transmitted on stigma (silk in maize) through the wind and insects. However, in self-pollinated plants like soybean, 10 anthers mature into flower buds (Figure 6.3) and pollen grains dehisce (Figure 6.4) prior to blooming of the flower (Figure 6.5). Pollen sterility is either controlled by genes (male-sterile, desynaptic or asynaptic genes—meiotic mutants) or chromosomal aberrations. Pollen sterility in F_1 obtained from distantly related interspecific and intergeneric hybrids is usually common. Hybrid plants differing by a reciprocal translocation usually express 50% pollen fertility, a key indicator that parents differ by a reciprocal translocation. The mature normal pollen grains are the result of normal meiosis.

6.2 PROTOCOLS FOR POLLEN FERTILITY

Pollen viability or fertility can be ascertained after staining the mature pollen with acetocarmine (1%), I_2-KI (potassium iodide; 1%), methylene blue and fuchsin, fluorescein diacetate, and 4', 6-diamidino-2-phenylindole (DAPI).

6.2.1 ACETOCARMINE STAIN

Determination of pollen fertility by staining pollen grains in acetocarmine is routinely used for cereal crops. This stain is quick and yields precise information on pollen fertility and sterility.

6.2.1.1 Fertile Pollen Grains
- Collect flowers during anthesis.
- Dust pollen grains on a clean slide. Add a drop of 1% acetocarmine, place a cover glass, and heat the slide over a low flame but be sure not to boil.
- Heating dissolves the starch grains and facilitates the staining of sperm nuclei and the vegetative nucleus.
- In barley, rye, maize, oats, and wheat, pollen grains with two well-developed sperm nuclei and one vegetative nucleus are considered functional pollen grains (Figure 6.1).

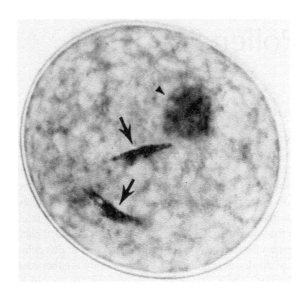

FIGURE 6.1   Mature normal fertile pollen grain of barley showing two sperm nuclei (arrow) and one vegetative nucleus (arrow head).

FIGURE 6.2   Field of maize plants showing exserted anthers on tassel branches.

FIGURE 6.3 By contras maize (Figure 6.2), soybean anthers mature in an early stage of flower bud and anthers are arranged in two rows surrounding the stigma.

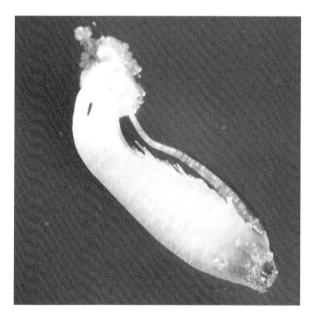

FIGURE 6.4 Pollen grains dehisced from the anthers; one filament with anther dissociated from the groups of 10.

FIGURE 6.5 Soybean flowers; gynoecium and androecium enclosed by keel ensuring self-pollination.

6.2.1.2 Sterile Pollen Grains

- Degenerated pollen grains are those containing two deformed sperm nuclei (Figure 6.6), one or two underdeveloped nuclei or empty pollen grains that are shrunken and have no cytoplasm and nucleus. Degenerated pollen grains exhibit various kinds of abnormalities (Figures 6.7 and 6.8). Figure 6.9 shows a shriveled and unstained pollen grain.

6.2.2 Iodine–Potassium Iodide (I_2–KI Stain)

In maize, this stain is used to distinguish waxy (tan color) and nonwaxy (purple) pollen grains (Figure 6.10: http://mutants.maizegdb.org/doku.php?id=waxy). Staining pollen grains with I_2–KI do not require heating, and fertile pollen grains turn black while sterile pollen grains are colorless.

- Dissolve 2.5 g KI + 250 mg I_2 in 125 mL ddH_2O.
- Dust pollens on a slide.
- Apply a drop of I_2–KI.
- Cover with a cover glass.
- Remove excess stain by a paper towel and observe under 20× magnification lens.

6.2.3 Alexander Stain

Alexander (1969) developed a stain for examining fertile and aborted pollen grains. The pollen staining depends upon the concentration and pH of stain and thickness of the pollen walls.

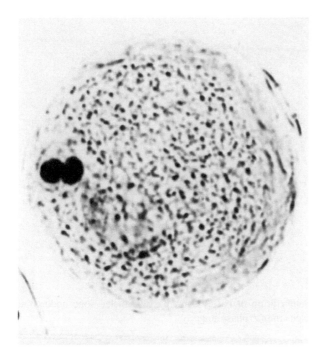

FIGURE 6.6   Pollen grain of barley showing two degenerated nuclei.

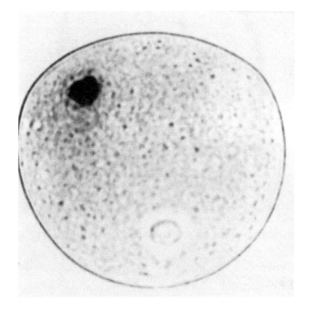

FIGURE 6.7   Pollen grain of barley with one stained nucleus and one showing vacuole-like structure.

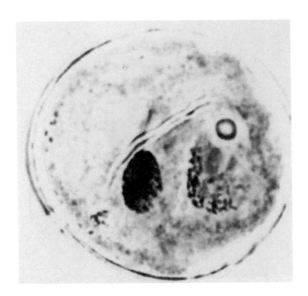

FIGURE 6.8 Pollen grain of barley showing two nuclei—one undergoing cell division while second nucleus in interphase stage.

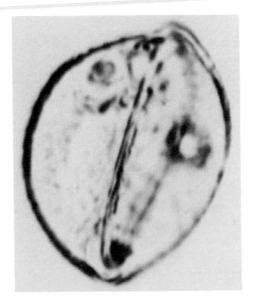

FIGURE 6.9 Shriveled unstained pollen grain considered sterile pollen.

6.2.3.1   Ingredients

10 mL	95% ethanol
10 mg	Malachite green (1 mL of 1% solution in 95% ethanol)
50 mL	dH$_2$O
25 mL	Glycerol

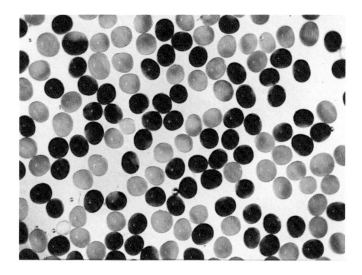

FIGURE 6.10 Maize pollen grains stained with I_2–KI segregating for tan (waxy) versus purple nonwaxy pollen grains (http://mutants.maizegdb.org/doku.php?id=waxy).

5 g	Phenol
5 g	Chloral hydrate
50 mg	Acid fuchsin (5 mL of 1% solution in dH_2O)
5 mg	Orange G (0.5 mL of 1% solution in dH_2O)
1–4 mL	Glacial acetic acid

Note: The amount of glacial acetic acid depends upon the thickness of the pollen walls: 1 mL (thin-walled pollen), 3 mL (thick-walled and spiny pollen), and 4 mL (nondehiscent anthers).

6.2.3.2 Staining

Sánchez-Morán et al. (2004) examined pollen fertility in *Arabidopsis thaliana* by using the Alexander (1969) stain. Stained fertile pollen grains of a wild type showed a magenta–red color (Figures 6.11 and 6.12), while sterile pollen grains did not stain red and were morphologically deformed (Figure 6.13).

Peterson et al. (2010) simplified the Alexander stain because the stain contains chloral hydrate and phenol. These chemicals are toxic and require a long time to acquire the optimum stain. They examined 22 plant species from the proposed staining technique.

- *Bud fixation*: fix flower buds or anthers prior to anthesis (mature anthers but not shedding pollen grains).
- Fix in 6:3:1 (ethanol:chloroform:glacial acetic acid) for a minimum of 3 h. Buds can be stored in fixative for 12 h at either RT or in a cold room.

6.2.3.3 Stain Solution

Ethanol (95%)	10 mL
Malachite green (1% solution in 95% ethanol)	1 mL
Distilled water	50 mL

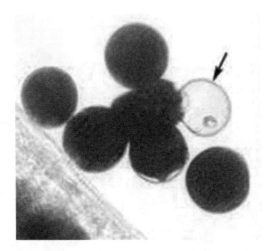

FIGURE 6.11 Pollen fertility in *Arabidopsis thaliana* by using Alexander (1969) stain. Stained fertile pollen grains of wild type showed magenta–red color, while sterile pollen grains did not stain red. (From Peterson, R., et al., *Intl. J. Plant Biol.*, 1, e13, 66–69, 2010.)

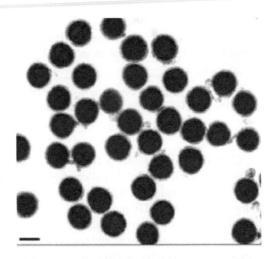

FIGURE 6.12 Pollen fertility in *Arabidopsis thaliana* by using Alexander (1969) stain; fertile pollen grains stained in magenta color. (From Sánchez-Morán, E., et al., *The Plant Cell*, 16, 2895–2909, 2004.)

Glycerol	25 mL
Acid fuchsin (1% solution in water)	5 mL
Orange G (1% solution in water)	0.5 mL
Glacial acetic acid	4 mL
Add dH$_2$O (4.5 mL) to a total of	100 mL

staining: Following at least 2 h post-fixation.

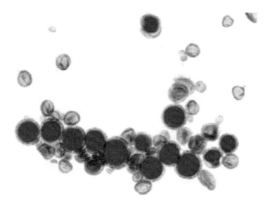

FIGURE 6.13 Pollen fertility in *Arabidopsis thaliana* by using Alexander (1969) stain; fertile pollen grains stained in magenta color and sterile pollen grains morphologically deformed and unstained. (From Sánchez-Morán, E., et al., *The Plant Cell*, 16, 2895–2909, 2004.)

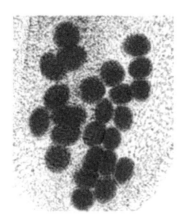

FIGURE 6.14 Modified Alexander stain showing fertile pollen grains in magenta–red of *Arabidopsis thaliana*. (From Peterson, R., et al., *Intl. J. Plant Biol.*, 1, e13, 66–69, 2010.)

- Place a bud on a microscope glass slide and remove excessive fixative using a filter paper.
- Apply 2–4 drops of stain.
- Dissect buds or anthers to release the pollen grains under a dissecting microscope.
- Remove debris and heat the slide slowly over an alcohol burner in a fume hood near the boiling stage. Slow heat helps the penetration of the stain into the cellulose and protoplast of the pollen.
- Place a cover glass over the sample and apply light even pressure.
- Seal the cover glass using nail polish and examine the slide using a light microscope.
  Fertile pollen grains show magenta–red pollen grains of *Arabidopsis thaliana* (Figure 6.14) and sterile pollen grains stain blue (Figure 6.15, arrow).

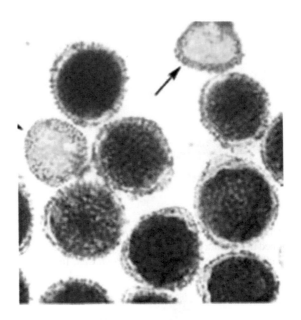

FIGURE 6.15 Fertile (magenta–red) and sterile (stained blue) pollen grains of *Arabidopsis thaliana* from modified Alexander stain. (From Peterson, R., et al., *Intl. J. Plant Biol.*, 1, e13, 66–69, 2010.)

6.2.4 Differential Staining of Pollen

6.2.4.1 Staining of Pollen with Thin Wall

- Thin-walled pollen occurs in *Phaseolus, Sorghum, Oryzae, Triticum, Hordeum, Zea,* and *Lycopersicon.*
- Stain pollen with one drop of stain, cover with a cover glass (slip), heat over a small flame, and examine under a microscope.
- If differentiation is not satisfactory, keep the slide in an oven at 50°C for 24 h.
- Store the mounted pollen for a week without fading of the stain.

6.2.4.2 Staining of Pollen with Thick and Spiny Walls

Nonsticky Pollen

- Acidify about 100 mL of the stain with 3 mL glacial acetic acid.
- Add a small quantity of pollen into a small vial and pour enough stain to cover the pollen.
- Keep in an oven at 50°C for 24–48 h.
- Examine after 24 h for differentiation.
- Add a drop of 45% acetic acid into the specimen vial and mix if aborted and nonaborted pollen are not well differentiated.

Sticky and Oily Pollen
- Acidify about 100 mL stain with 3 mL glacial acetic acid.
- Fix mature but nondehisced anther for 24 h in a fixative (3 ethanol:2 chloroform:1 glacial acetic acid). This fixative removes sticky materials.
- Transfer through an ethanol-water series to water.
- Remove the excess water by placing anthers between filter papers.
- Cover one or two anthers with one to two drops of stain on a slide, split the anther wall with a needle to release the pollen, and remove debris.
- Place a cover slip over the stain and store in an oven at 50°C for 24–48 h.
- After 24 h, add one to two drops of stain along the sides of the cover slip to replace the amount of stain lost during evaporation.
- Examine the slide after 24 h. If the green color dominates, increase the acidity of the stain mixture. This stain is suitable for *Hibiscus* and *Cucurbita*.

6.2.4.3 Staining of Pollen inside a Nondehiscent Anther
- Acidify 100 mL stain with 4 mL glacial acetic acid.
- Collect anthers immediately after anthesis.
- Fix for 24 h in a fixative:

60 mL	Methanol
30 mL	Chloroform
20 mL	dH_2O
1 g	Picric acid
1 g	$HgCl_2$

- Transfer through 70%, 50%, and 30% ethanol, 30 min each; change gradually to hydrate the anthers, and finally rinse with dH_2O.
- Remove excess water between two filter papers.
- Stain anthers and keep in an oven at 50°C for 24–48 h.
- Remove excess stain with blotting paper and examine. If overstained, remount in 25% glycerol containing 4% chloral hydrate, and seal.

6.3 POLLEN–STIGMA INCOMPATIBILITY

6.3.1 POLLEN GERMINATION TEST

To ascertain compatibility in wide crosses, pollen germination, pollen tube growth, and fertilization are determined by fluorescence microscopy.

The pollen germination protocol developed by Rodriguez-Enriquez et al. (2013) for *Arabidopsis thaliana* requires the following medium. This medium was derived with several formulations.

- Agarose 0.5%
- 0.01% boric acid
- 1 mM $CaCl_2$
- 1 mM $Ca(NO_3)_2$

- 1 mM KCl
- 0.03% casein enzymatic hydrolysate
- 0.01% myoinositol
- 0.01% ferric ammonium citrate
- 10 mM gamma-aminobutyric acid (GABA)
- 0.25 mM spermidine
- 10% sucrose
- pH 7.0

6.3.1.1 Method

- Dissolve sucrose first and then add the other ingredients; adjust pH to 8.0.
- Finally add agarose; mix and briefly heat in the microwave until it is dissolved.
- Keep the medium in a water bath (c. 40°C–65°C).
- *For germination and pollen tube growth*: Layer agarose pad onto an uncoated glass microscope slide.
- Place slides on a completely flat surface. Add 500 μL hot solution to form a flat agarose pad.
- Store slides vertically in a microscope slide container with wet tissue paper placed in the container (but not touching the slide surface) to ensure a constant saturated humidity.
- Seal the container and keep at 4°C; slides can be used for up to 7 days following preparation.

6.3.1.2 Pollen Culture

- Use freshly made slides at RT (cold storage slides can also be used).
- To maximize pollen tube growth, modification in osmoticum (sucrose) and spermidine is done.
- Keep plate with pollen grains at 28°C and pH 8.0.

6.3.1.3 Humidity Chamber

- Prepare a humidity chamber by placing a layer of tissue paper wetted with warm (~40°C–50°C) water at the bottom of a large square Petri dish.
- Place a glass rod on top of the paper to support the slide.
- Equilibrate the slide in the chamber for 10–15 min.
- Cut rectangular pieces (~1.2 cm × 1.7 cm), always smaller (1.2 cm × 1.7 cm), always smaller than the pad surface.
- Place on the top of agarose pad with the help of two forceps, leaving an edge of agarose pad on all sides.
- Place pollen on the surface of the cellophane membrane.
- Select a freshly opened flower in the morning and dust pollen grains.
- Immediately place slides in a vertical position in the same slide box.
- Seal the box and incubate the slides in the dark at 24°C.

6.3.1.4 Microscopic Examination

- Examine (count) pollen germination using random field.
- Examine about 300 pollen grains and score germinated or ungerminated.
- *Pollen tube structure*: Remove pollen tube from the cellulose layer following 5 h in culture.
- Add a small amount of sucrose solution onto the membrane to form a small pool.
- Quickly remove pollen tubes floating on the pool surface and keep them on the surface of the slide.
- Fix pollen tube on the slide in 3:1 ethanol:glacial acetic acid for 10 min.
- Transfer to 70% ethanol and rehydrate in distilled water for 15 min.
- Drain and stain in aniline blue and mount in DABCO; 2 mg mL^{-1} (Sigma) containing DAPI and 100 ng mL^{-1}. DAPI stock solution dissolved in glycerol containing 10% 1 M Tris buffered to pH 8).
- Examine pollen tubes using epifluorescence (Figure 6.16).

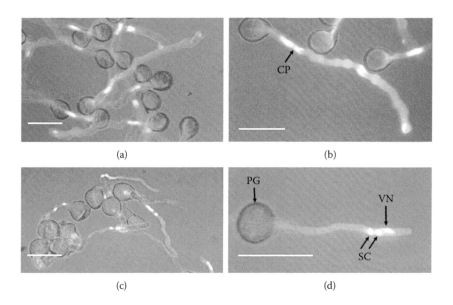

(a)   (b)

(c)   (d)

FIGURE 6.16 Pollen germination test to determine pollen viability; growth of pollen tubes determined by using epifluorescence microscope. Pollen tube growth of *Arabidopsis thaliana* Col-0 *in vitro*. Pollen tubes are shown following 5 h culture at 24°C on a cellulosic membrane overlying a 0.5% agarose medium with 18% sucrose, 0.01% boric acid, 1 mM CaCl$_2$, 1 mM Ca(NO3)$_2$, 1 mM KCl, 0.03% casein enzymatic hydrolysate, 0.01% myoinositol, 0.01% ferric ammonium citrate, and 0.25 mM spermidine, pH 8. (a, b) Tubes stained with aniline blue to show callose plugs. (c, d) Pollen tubes stained with DAPI revealing the location of sperm cells (generally spherical) and the vegetative nucleus (more diffuse and elongated). Arrows indicate examples of key features: PG, pollen grain; CP, callose plug; SC, sperm cell; VN, vegetative nucleus. Bars, 50 μm.

Fang et al. (2010) examined pollen viability by germination test in chickpea. The
protocol is:

- Collect pollens in an Eppendorf tube by squeezing the keel from the base
  upward with forceps until pollens are exuded through the tip.
- Dissolve fluorescein diacetate (FDA, 2 ng) in 1 mL of acetone.
- Put one drop of the acetone–FDA solution on a microscope slide and allow
  to evaporate.
- Mix pollen with a 10% sucrose solution.
- Place a drop of the solution on the evaporated acetone–FDA and cover with
  a cover glass.
- Examine pollen grains with gray color (unviable pollen) under a fluores-
  cence microscope (Figure 6.17a through 6.17d).

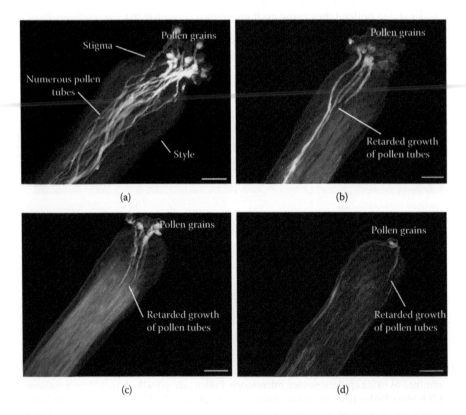

FIGURE 6.17   Pollen tube growth down the style in pistils of (a) well-watered (WW) Rupali
chickpeas pollinated with pollen from WW plants (WW+WW); (b) WW plants pollinated
with pollen from water-stressed (WS) plants (WW+WS); (c) WS plants pollinated with pollen
from WW (WW+WS); and (d) WS plants pollinated with pollen from WS plants (WS+WS)
in Experiment 2. Styles were harvested when the predawn leaf water potential of WS plants
was 1.2 MPa and then fixed 24 h later and stained with aniline blue. Scale bar ¼ 50 lm. (From
Fang et al., *Jour. Exptl. Bot.* 61, 335–345, 2010.)

- Inoculate pollen grains into 100 mL of freshly prepared pollen culture medium of Brewbaker and Kwack (1963) at 25°C:
  10% sucrose
  100 ppm H_3BO_3
  300 ppm $Ca(NO_3)_2.4H_2O$
  200 ppm $MgSO_4.7H_2O$
  100 ppm KNO_3
  Dissolve in distilled water
- Mix the contents on a vertex mixer and incubate in the dark.
- After 4 h, add one drop of fixative (3 ethanol:1 glacial acetic acid; v/v) to the sample.
- Estimate pollen fertility by examining 300 pollen grains (10–15 microscopic fields).
  Germination of small pollen populations (five or more tests) of *Ornithogalum virens* pollen in 1/50 mL drops. It appears that coconut milk in 10:100 produced 60.2% pollen germination.

Culture medium	% germination
10% sucrose + 10 ppm boric acid (10:100)	0
Pollen extract in 10:100	42.9
Plant tissue extract[a] in 10:100	50.1
Coconut milk in 10:100	**60.2**
Yeast extract in 10:100	36:5
Amino acid in 10:100	0
Vitamins in 10:100	0
Purines and pyrimidines in 10:100	0
Pollen-ash extract in 10:100	48.5
Pollen-ash extract run through cation exchange resin, in 10:100	0

[a] Tissues include petals, pistils, stems, leaves, ovaries, and roots.

6.3.2 STAINING PISTILS WITH ANILINE BLUE

To ascertain compatibility in wide crosses, pollen germination, pollen tube growth, and fertilization are determined by fluorescence microscopy (Kho and Baër 1968).
Fixative: Fix gynoecia in 24 h postpollination.
1 part formaldehyde (HCHO approx. 37%)
1 part propionic acid
18 parts 70% ethanol

- After 24 h postpollination, wash gynoecia once with ddH_2O.
- Transfer gynoecia in 1N NaOH at RT (25°C) for 24 h to soften the tissues.
- Wash softened gynoecia twice, 2 min each, with ddH_2O and stain for 24 h in 0.1% water soluble aniline blue solution prepared in 0.1N K_3PO_4.

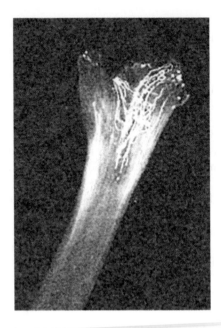

FIGURE 6.18 Pollen tube growth in a self-compatible *Lycopersicon peruvianum* 16 h post-pollination; pollen tubes are growing through style. (From Kho, Y.O., and Baër, J., *Euphytica*, 17, 298–302, 1968.)

- Squash gynoecia gently in a drop of 80% glycerin under a cover glass on a clean slide and observe using a fluorescence microscope.
- Pollen tube growth in a self-compatible *Lycopersicon peruvianum* 16 h postpollination is shown in Figure 6.18.
- Pollen tube growth is arrested in the style, 16 h postpollination in an incompatible pollination (Figure 6.19).
- In a compatible cross of *G. clandestina* × *G. max*, the pollen tube had already reached the ovules 24 h postpollination (Figure 6.20).

6.3.3 Cleared-Pistil Method

Young et al. (1979) developed a protocol known as cleared-pistil and thick-sectioning technique for detecting apomixis in grasses.

- Collect inflorescences either from field-grown or from greenhouse-grown plants at the time of stigma exertion.
- Fix in what is known as FAA (v/v):

  1. Formalin = 3
  2. Glacial acetic acid = 3
  3. Ethanol (95%) = 40
  4. Water: 14

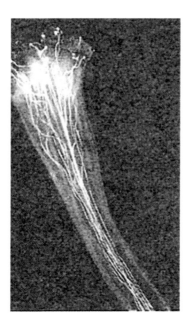

FIGURE 6.19 Pollen tube growth is arrested in the style, 16 h postpollination in an incompatible pollination in *Lycopersicon peruvianum*. (From Kho, Y.O., and Baër, J., *Euphytica*, 17, 298–302, 1968.)

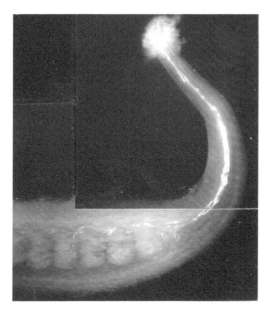

FIGURE 6.20 In a compatible cross of *Glycine clandestina* x *G. max*, pollen tube had already reached the ovules 24 h postpollination. (From Singh, R.J., and Hymowitz, T., *Plant Breed.*, 98, 171–173, 1987.)

- Excise fixed pistils and place in 50% ethanol in a 15 mm × 60 mm screw-cap vial.
- Dehydration and clearing step:
  70% ethanol →85% ethanol →100% ethanol (three changes)
    1 (ethanol):1 (methyl salicylate) → 1 (ethanol):3 (methyl salicylate) →
  100% methyl salicylate (two changes).

Note: Place one excised pistil in a separate vial with 1 mL liquid for 30 min. Later, wrap 10–20 pistils from an inflorescence in a moist envelope from 12 mm × 12 mm square of single-thickness Kimwipes tissue, place in a vial, and clear using 2 mL solution for 2 h each step.

- Change the solution with a Pasteur pipette, store cleared pistils in methyl salicylate in vials and examine with an interference contrast microscope.
- Mount cleared pistils in methyl salicylate under an unsealed cover slip on a microscope slide.
- **Record components** of embryo sac and photograph with a camera attached with the computer.
- After examination by interference contrast microscopy, transfer each pistil to a vial with 50% ethanol for embedding in plastic.
- The infiltration steps are as follows:
    50% ethanol → 75% ethanol → 100% ethanol (three changes) → 100% propylene oxide (two changes) → 50% Spurr's low viscosity embedding medium → 100% Spurr's (four changes). Steps through 50% Spurr's are 30 min each and after 50% Spurr's are 1 h each. At each Spurr's medium step, agitate the vials briefly on a vortex mixture and evacuate.
- Spurr's low viscosity embedding medium (Spurr 1969):
    - Epoxy resin (ERL 4206: vinyl cyclohexene dioxide MW 140.18)   10 g
    - Flexibilizer (DER-736: curing agent (low viscosity),
      an epoxide MW 175–205, av. 380)   6 g
    - NSA (nonenylsuccinic anhydride)   26 g
    - S-1 or DMAE (dimethylaminoethanol, curing agent)   0.2g
- Cure schedule 16 h at 70°C.
- Remove styles after the final changes, align ovaries in a size 00, square-tipped BEEM capsules, and cure the resin for 16 h at 70°C.
- For sectioning, coat untrimmed blocks on the top and bottom edges with a mixture of Elmer's contact cement and toluene (1:2 v/v), dry overnight at RT and section at 1.8 μm with a glass knife on a Sorvall MT2B ultramicrotome.
- Collect ribbons, 6–10 sections long, on a Teflon-coated slide and then transfer to a large drop of freshly boiled dH_2O on a slide coated with 0.5% gelatin. Expand the ribbons and dry on a hot plate at 65°C.
- Heat slide carefully for 20 s over an alcohol flame to secure the sections.
- Stain sections by floating the slide with 1% toluidine blue O in 0.05 M borate buffer at pH 9 for 4–8 min at 65°C.
- Rinse slide briefly with each of dH_2O, 95% ethanol, 100% ethanol, and xylene.
- Affix a cover slip with Permount.

6.3.4 Modified Cleared-Pistil Method

The following modified cleared-pistil method was developed by Crane and Carman (1987) and C.B. do Vale (personal communication, 2002). Pistils should remain in each of the following solutions at least for 30 min:

- Fix inflorescence in 6:3:1 (v/v/v/) 95% ethanol:chloroform:glacial acetic acid for 24–48 h.
- Store in 75% ethanol at −15°C.
- Transfer pistils in 95% ethanol → 2:1 (v/v) 95% ethanol: benzyl benzoate → 1:2 (v/v) 95% ethanol: benzyl benzoate → 2:1 (v/v) benzyl benzoate:dibutyl phthalate (teratogen) at intervals greater than 1 h clearing solution.
  Remove solution carefully from pistil-containing vials by a fine bent Pasteur pipette. Some solution may remain with the pistil, but remove as much as possible without sucking the pistil into the pipette.
- Place cleared pistils on a microscope slide in a sagittal optical section between two cover slips.
- Fill spaces with the clearing solution.
- A third cover slip is then placed on top, resting on the two other cover slips. Avoid formation of air bubbles.

Note: Time in clearing solution may be species specific. *Brachiaria* requires a longer time than *Panicum*.

CAUTION: Avoid contact with skin!

6.4 CHROMOSOME COUNT IN POLLEN

Kindiger and Beckett (1985) developed stains to count chromosomes in pollen grains of maize.

6.4.1 Stain

Solution A: Dissolve 2 g hematoxylin in 100 mL 50% propionic acid. Allow the solution to age for about 1 week. The solution can be kept indefinitely in a stoppered brown bottle at RT.

Solution B: Dissolve 0.5 g ferric ammonium sulfate [$FeNH_4 (SO_4)$] in 100 mL 50% propionic acid. The solution can be kept indefinitely in a stoppered brown bottle at RT.

Note: Mix equal volumes of solution A and B. The mixture, which turns dark brown, is ready for use immediately. The stain remains good for about 2 weeks.

The following protocol was developed to count chromosomes from maize pollen grains:

- Collect fresh maize tassels (when about 3 cm of the tassel becomes visible above the leaf whorl) in a waxed short bag or other moisture-retaining container.

- Fix tassel in a 14:1 mixture of 70% ethanol:formaldehyde and store in a refrigerator. This fixative produces better results than 3:1 (95% ethanol:glacial acetic acid).
- Pretreatment is unnecessary for the fresh material. However, anthers can be pretreated with α-bromonaphthalene with a drop of DMSO for 30–40 min; fix anthers in glacial acetic acid.
- Dissect out anthers and place in a drop of 45% acetic acid on a clean slide.
- Add chloral hydrate crystal, remove debris, add a drop of stain, macerate, and stain for 1 min. Apply cover slip, heat (but not boil), and remove excess stain with a filter paper by applying gentle pressure. Heat helps to darken the chromosomes and nuclei and clears the cytoplasm (Figures 6.21 through 6.25).
- To make the slide permanent, seal the edges of the cover slip with Permount and allow to air-dry. Stain begins to fade after 3 months.

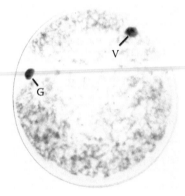

FIGURE 6.21 Vegetative (V) and generative (G) nuclei. (From Kindiger, B., and Beckett, J.B., *Stain Technol.*, 60, 265–269, 1985.)

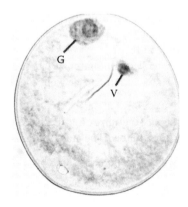

FIGURE 6.22 Initiation of mitosis in the generative nucleus. (From Kindiger, B., and Beckett, J.B., *Stain Technol.*, 60, 265–269, 1985.)

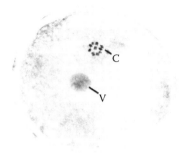

FIGURE 6.23 Metaphase chromosome (*n* = 10). (From Kindiger, B., and Beckett, J.B., *Stain Technol.*, 60, 265–269, 1985.)

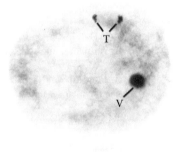

FIGURE 6.24 Telophase stage in the generative nucleus. (From Kindiger, B., and Beckett, J.B., *Stain Technol.*, 60, 265–269, 1985.)

FIGURE 6.25 Generative nucleus produces two sperm nuclei after mitosis. (From Kindiger, B., and Beckett, J.B., *Stain Technol.*, 60, 265–269, 1985.)

REFERENCES

Alexander, P. 1969. Differential staining of aborted and nonaborted pollen. *Stain Tech.* 44: 117–122.

Brewbaker, J. L., and B. H. Kwack. 1963. The essential role of calcium ion in pollen germination and pollen tube growth. *Amer. J. Bot.* 59: 859–865.

Crane, C. F., and J. G. Carman. 1987. Mechanism of apomixis in *Elymus rectisetus* from eastern Australia and New Zealand. *Amer. J. Bot.* 74: 477–496.

Fang, X., N. C. Turner, G. Yan, F. Li, and K. H. M. Siddique. 2010. Flower numbers, pod production, pollen viability, and pistil function are reduced and flower and pod abortion increased in chickpea (Cicer arietinum L.) under terminal drought. *J. Expt., Bot.* 61: 335–345.

Kho, Y. O., and J. Baër. 1968. Observing pollen tubes by means of fluorescence. *Euphytica.* 17: 298–302.

Kindiger, B., and J. B. Beckett. 1985. A hematoxylin staining procedure for maize pollen grain chromosomes. *Stain Technol.* 60: 265–269.

Peterson, R., J. P. Slovin, and C. Chen. 2010. A simplified method for differential staining of aborted and non-aborted pollen grains. *Int. J. Plant Biol.* 1: 66–69.

Rodriguez-Enriquez, M. J., S. Mehdi, H. G. Dickinson, and T. T. Grant-Downton. 2012. A novel method for efficient in vitro germination and tube growth of Arabidopsis thaliana pollen. *New Phyto.* 197: 668–679.

Rodriguez-Enriquez, M. J., S. Mehdi, H. G. Dickinson, and R. T. Grant-Downton. 2013. A novel method for efficient in vitro germination and tube growth of *Arabidopsis thaliana* pollen. *New Phyt.* 197: 668–679.

Sánchez-Morán, E., G. H. Jones, F. C. H. Franklin, and J. L. Santos. 2004. A Puromycin-sensitive aminopeptidase is essential for meiosis in *Arabidopsis thaliana. The Plant Cell.* 16: 2895–2909.

Singh, R. J., and T. Hymowitz. 1987. Intersubgeneric crossability in the genus Glycine Willd. *Plant Breed.* 98: 171–173.

Spurr, A. R. 1969. A low-viscosity epoxy resin embedding medium for electron microscopy. *J. Ultrastruct. Res.* 26: 31–43.

Young, B. A., R. T. Sherwood, and E. C. Bashasw. 1979. Cleared-pistil and thick-sectioning techniques for detecting aposporous apomixis in grasses. *Can. J. Bot.* 57: 1668–1672.

7 Cell Division

7.1 INTRODUCTION

- Cell division is a continuous process that occurs in all living organisms.
- It has been divided into two categories: mitosis and meiosis.
- Both forms of nuclear division occur in eukaryotes and these process comprise the cell cycle: G_1 (growth) $\rightarrow$ S (synthesis of DNA) $\rightarrow$ G_2 (growth) $\rightarrow$ M (mitosis or meiosis) $\rightarrow$ C (cytokinesis) (Smith and Kindfield 1999).
- Mitosis occurs in somatic tissues where each chromosome is divided identically into halves, both qualitatively and quantitatively, producing genetically identical to the parent nucleus.
- In contrast, meiosis takes place in germ cells with the consequence that nuclei with haploid chromosome numbers are produced.
- Both types of cell division play an important role in the development and hereditary continuity of a eukaryotic organism.

7.2 MITOSIS

7.2.1 PROCESS OF MITOSIS

- The term mitosis is derived from the Greek word *mitos* for thread; coined by Flemming in 1879 (Singh 2017).
- The synonym of mitosis is karyokinesis, that is, the actual division of a nucleus into two identical parental daughter nuclei.
- It is also known as equational division because the exact longitudinal division of each chromosome into identical chromatids and their precise distribution into daughter nuclei leads to the formation of two cells—identical to the original cell from which they were derived.
- The process of mitotic cell division has been divided into six stages: (1) interphase, (2) prophase, (3) metaphase, (4) anaphase, (5) telophase, and (6) cytokinesis.

7.2.1.1 Interphase

- Two more terms, resting stage and metabolic stage, have been used to identify interphase cells. However, interphase cells should not be described as being in a "resting stage" because their nuclei are very active as they prepare for cell division.
- The DNA replication and transcription occur during interphase (Manuelidis 1990). Interphase consists of three phases: G_1 (gap 1; pre-DNA synthesis) phase, S phase (DNA synthesis), and G_2 (gap 2; post-DNA synthesis).

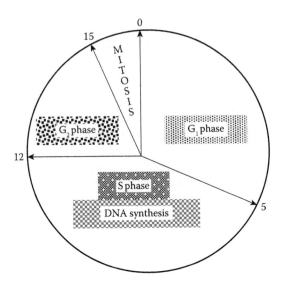

FIGURE 7.1 Mitotic cycle.

- The duration of mitotic division is short compared to time required for the cells going through interphase (Figure 7.1). Thus, "metabolic stage" is a more appropriate term for the interphase cells.
- The interphase nucleus contains one or more prominent nucleoli and numerous chromocenters depending upon the heterochromatic nature of the chromosomes. Chromosomes cannot be traced individually and they are very lightly stained (Figure 7.2a).

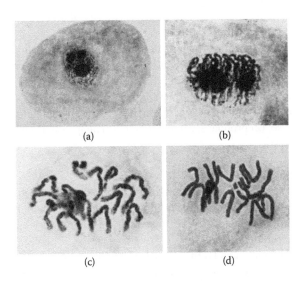

FIGURE 7.2 (a–d) Mitosis in barley ($2n = 14$) without pretreatment and with acetocarmine staining. (a) Interphase, (b) early prophase, (c) late prophase, and (d) metaphase.

7.2.1.2 Prophase

- All the chromosomes are beginning to become distinct and they are uniformly distributed in the nucleus. In early prophase, chromonemata become less uniform.
- The chromosomes are more or less spirally coiled and seem to be longitudinally double (Figure 7.2b).
- The two longitudinal halves of a chromosome are known as chromatids. As the prophase stage advances to mid and late prophase, chromosomes become thicker, straighter, and smoother (Figure 7.2c).
- The two chromatids of a chromosome become clearly visible. The nucleoli begin to disappear in late prophase.
- It was formerly believed that chromosomes in a prophase nucleus are arranged in a haphazard fashion throughout the nucleus, but this is not true.
- Several studies have shown that in interphase and prophase nuclei, kinetochores (centromeres) or primary constrictions are oriented toward one pole while telomeres face opposite to the kinetochores and are attached to the nuclear membrane.
- This orientation suggests that chromosomes substantially maintain the previous telophasic position (Figure 7.2b, c).

7.2.1.3 Metaphase

- During metaphase, kinetochores move to the equatorial plate.
- Nucleoli and nuclear membrane disappear.
- Chromosomes are shrunk to the minimum length. Kinetochores are attached to spindle fibers, while chromosome arms may float on either side in the nucleus (Figure 7.2d). Karyotype analysis of a species is generally studied at metaphase after pretreatment of specimen cells.

7.2.1.4 Anaphase

- As the anaphase stage ensues, kinetochores become functionally double and the two sister chromatids of each chromosome separate (Figure 7.3a).
- It appears that the kinetochores (spindle fiber attachment) regions of the two sister chromatids are being pulled to opposite poles by the spindle fibers.
- Chromatids are more slender and densely stained. Each daughter chromosome moves to a polar region. (Figure 7.3b).
- By the end of anaphase, spindle fibers disappear and the compact groups of chromosomes at the two poles are of identical genetic constitution.

7.2.1.5 Telophase

- Chromosomes contract and form a dense chromatid ball.
- Chromosomes of the two daughter nuclei reorganize at the telophase.
- Nucleoli, nuclear membranes, and chromocenters reappear, and the chromosomes lose their stainability (Figure 7.3c).

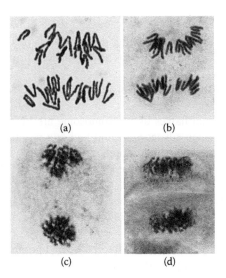

FIGURE 7.3   (a–d) Mitosis in barley ($2n = 14$) without pretreatment and with acetocarmine staining. (a) Anaphase, (b) late anaphase, chromosomes have already reached their respective poles, (c) telophase, and (d) cytokinesis.

7.2.1.6 Cytokinesis

- The division of cytoplasm and its organelles between daughter cells is called cytokinesis, which begins during the late telophase stage at or near the equatorial (metaphase) plate (Figure 7.3d).
- At mitotic anaphase, chromosomes remain localized during interphase and eventually reappear in the same position (kinetochores and telomeres are located at opposite sides in the nucleus during next prophase) (Roble model) (Figures 7.4a, b and 7.5a, b).

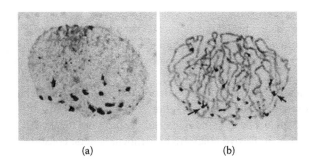

FIGURE 7.4   (a and b) Mitotic nuclei of rye ($2n = 14$) after Giemsa C-banding staining. (a) Interphase, with chromocenters (telomeric bands) on the lower portion of nucleus and spindle pole (kinetochores) on the upper side of the nucleus. (b) An early prophase nucleus with fused telomeres (arrows). (From Singh, R.J. and G. Röbbelen, Z. *Pflanzenzüchtg,* 75, 270–285, 1975. With permission.)

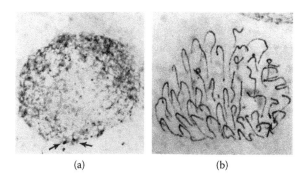

(a)   (b)

FIGURE 7.5   (a and b) Mitotic nuclei of wheat-rye ditelo-addition line ($2n = 42$ wheat + 2 telo rye) after Giemsa C-banding staining. (a) Interphase with two chromocenters (arrows) representing telomeres of the rye telocentric chromosomes. (b) An early prophase with dark telomeric bands (arrow) showing somatic association. (From Singh, R.J. et al. *Chromosoma (Berlin)*, 56, 265–273, 1976. With permission.)

- Cytokinesis in plants differs from that in animals. In plants, it takes place by the formation of a cell plate but in animals, cytokinesis begins by furrowing.

7.3 MEIOSIS

7.3.1 Process of Meiosis

7.3.1.1 Cycle 1

1. *Prophase-I*
   a. Leptonema
      - The chromonemata become more distinct from one another at leptonema and appear as very long and slender threads.
      - The chromosomes may be oriented with kinetochores toward the same side of the nucleus, forming a so-called bouquet.
      - Leptotene cells have large nucleoli and distinct nuclear membranes (Figure 7.6a).
   b. Zygonema
      - The mechanism of synapsis of homologous chromosomes is a complex phenomenon and is described by light microscope, electron microscope (synaptonemal complex-SC), and molecular experiments. The light microscope view is presented here.
      - Synapsis starts initially from one or more regions of homologous chromosomes and gradually extends at several secondary sites zipper-like until it is complete (Burnham et al. 1972).
      - Chromosomes begin to shorten and thicken. (Figure 7.6b). A physical multiple interstitial interhomologue connection is a prerequisite for homologous chromosome pairing and recombination (Kleckner 1996).

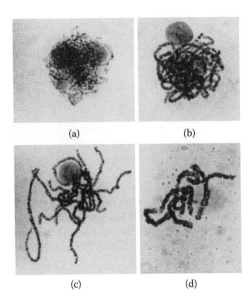

(a)   (b)

(c)   (d)

FIGURE 7.6 (a–d) Meiosis in barley ($2n = 14$) after propionate carmine staining. (a) Leptonema, (b) zygonema, (c) pachynema, and (d) an early diplonema.

  - The site of chromosome pairing initiation is still debated. The intimate association of homologous chromosomes in the synaptonemal complex, alignment of homologous chromosomes, and recombination has been critically reviewed by Santaos (1999). Recombination should be required for synapsis.
c. Pachynema
  - The synapsis of homologous chromosomes is completed. Chromosome regions unpaired in pachynema will remain usually unpaired.
  - Chromosomes are noticeably thicker and shorter than those in the leptonema, and paired entities visibly represent the haploid number of that species.
  - Each pair of chromosomes is termed a bivalent. The term bivalent is used for the paired homologous chromosomes.
  - Some scientists still use an outdated term tetrad. Tetrad should be used only for the four meiotic products such as the tetrad after meiosis in anthers.
  - The unpaired chromosome is a univalent and three paired homologous chromosomes are trivalent, etc.
  - Nucleoli are clearly visible and certain chromosomes may be attached to them. These chromosomes are known as nucleolus organizer chromosomes or satellite (SAT) chromosomes (Figure 7.6c).

    d. Diplonema
        – Chromatids composing bivalents begin to separate, or repel, at one
          or more points along their length (Figure 7.6d). Each point of con-
          tact is known as a chiasma (pl, chiasmata).
        – At each point of contact, two chromatids have exchanged portions
          by crossing over. The number of chiasmata varies from organism
          to organism.
        – In general, longer chromosomes have more chiasmata than shorter
          chromosomes.
        – Chiasmata may be interstitial or terminal.
        – Interstitial chiasmata may be found anywhere along the length of a
          chromosome arm but terminal chiasmata are located at a chromo-
          some tip (telomere).
        – In most cases a terminal chiasma occupies an interstitial position
          earlier but as cell division progresses, its position moves to the tip
          of the chromosome arm in which it occurred (Figure 7.7a). This
          phenomenon has been termed terminalization.
    e. Diakinesis
        – Chromosomes continue to shorten and thicken.
        – In a squash preparation, compact and thick chromosomes lie well-
          spaced in the nucleus, often in a row near the nuclear membrane.
        – This is a favorable stage to count the chromosomes. The number of
          chiasmata is reduced due to terminalization.
        – The nucleolus begins to decrease in size (Figure 7.7b).
2. *Metaphase-I*
    a. As soon as cells reach the metaphase stage, the nuclear membrane and
       nucleoli disappear, and spindle fibers appear.
    b. Bivalents are arranged at the equatorial plate with their kinetochores
       facing the two poles of the cell.
    c. Chromosomes reach their maximum contractions (Figure 7.8a).

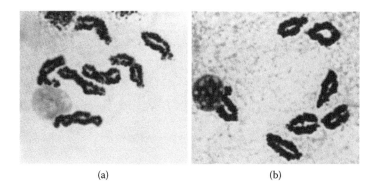

(a)   (b)

FIGURE 7.7 (a and b) Meiosis in barley ($2n = 14$) after propionate carmine staining. (a) Late
diplonema, showing 7 bivalents. (b) Diakinesis, a bivalent associated with nucleolus.

3. *Anaphase-I*
   a. Homologous chromosomes begin to separate toward the opposite poles of the spindle (Figure 7.8b).
   b. The dissociation of chromosomes initiates on schedule leaving an unpaired chromosome behind.
   c. Chromosome movement to their respective poles is either coupled with the shortening of spindle microtubules or kinetochore motors chew microtubules as they drag chromosomes to the poles (Zhang and Nicklas 1996).
   d. This generates two groups of dyads with the haploid chromosome number (Figure 7.8c).
   e. The meiotic anaphase-I chromosomes are much shorter and thicker than the chromosomes of mitotic anaphase.
4.. *Telophase-I*
   a. At the end of anaphase-I, chromosomes reach their respective poles and polar groups of chromosomes become compact (Figure 7.8d).
   b. Nuclear membranes and nucleoli start to develop and eventually two daughter nuclei with haploid chromosomes are generated.
   c. The chromatids are widely separated from each other and show no relational coiling.
   d. As a result of crossing over, each chromatid derived from a particular bivalent may be different genetically from either of the parental homologs that entered the bivalent.

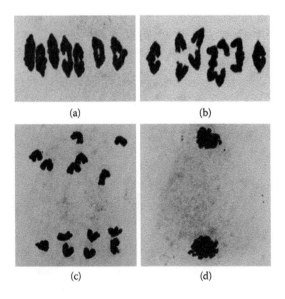

(a)   (b)

(c)   (d)

FIGURE 7.8 (a–d) Meiosis in barley ($2n = 14$) after propionate carmine staining. (a) Metaphase-I showing 7 bivalants. (b) An early anaphase-I, chromosomes are beginning to disjoin. (c) Anaphase-I showing 7–7 chromosome migration. (d) Telophase-I.

5. *Interkinesis*
   a. Interkinesis is the time gap between division I and division II.
   b. Generally it is short or may not occur at all. At the end of telophase-I, cytokinesis does not invariably follow division I.
   c. In many vascular plants two daughter nuclei lie in a common cytoplasm and undergo second division.

7.3.1.2 Cycle 2

- In the second division of meiosis, prophase-II (Figure 7.9a), metaphase-II (Figure 7.9b), anaphase-II (Figure 7.9c), and telophase-II, resemble similar stages of mitotic divisions.
- This second division yields a quartet (tetrad) of uninucleate cells (Figure 7.9d) called microspores (in male) produced from pollen mother cells (PMC) and megaspores (in female) produced from megaspore mother cells (MMCs).
- Tetrad analysis is a powerful scientist's tool to determine gamete development, cell division, chromosome dynamics, and recombination (Copenhaver et al. 2000).

7.4 GAMETOGENESIS

- The end products of meiosis in higher plants are four microspores in the male and four megaspores in the female containing the haploid (n) chromosome number.

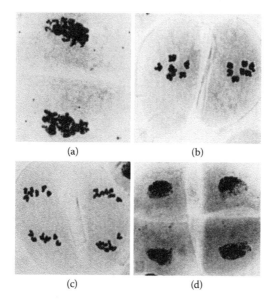

(a)   (b)

(c)   (d)

FIGURE 7.9 (a–d) Meiosis in barley ($2n = 14$) after propionate carmine staining. (a) Prophase-II, (b) metaphase-II, (c) anaphase-II, and (d) a quartet or tetrad cell.

- Two mitotic divisions occur in microspores. The first division produces a tube nucleus and a generative nucleus. The second division occurs only in the generative nucleus, producing two sperm cells (Figure 7.10).
- Megaspores are produced in the ovules from a MMC. Of the four megaspores produced by meiosis, only one normally survives.
- Three mitotic divisions occur during megagametogenesis. In the first division, a megaspore nucleus divides to give rise to the primary micropylar and primary chalazal nuclei.
- The second mitosis produces two nuclei each at the micropylar and at the chalazal region of the embryo sac. The third mitosis results in four nuclei at each of the opposite poles of an embryo sac.
- One nucleus from each pole, known as polar nuclei, moves to the middle of the embryo sac and the two nuclei fuse to give rise to a secondary nucleus, or polar fusion nucleus, with $2n$ chromosome constitution.
- At the micropylar end of the embryo sac, of the three nuclei remaining one nucleus differentiates into an egg cell and the remaining two become synergids (member of the egg apparatus).
- The three nuclei at the chalazal end of the embryo sac are called antipodal nuclei (Figure 7.10). Cell wall formation around the antipodals, synergids, and egg occurs. At maturity, the embryo sac consists of following cells: antipodals (3), synergids (2), egg (1), and the large central cell that contains the polar nuclei.
- There are many variations to embryo sac development detailed in Maheshwari (1950). The present text covers a typical gametogenesis of many plants, particularly the grasses so important in plant breeding.

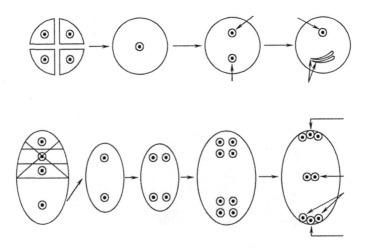

FIGURE 7.10 Diagrammatic explanation of gametogenesis in higher plants. Formation of pollen grains (top). Development of embryo sac (bottom); three antipodal nuclei (top), two polar nuclei (center); three nuclei (egg) are surrounded by synergids.

7.5   FERTILIZATION

- During pollination and fertilization in higher plants, pollen grains fall onto the stigma.
- The tube nucleus of a pollen grain directs the growth of the pollen tube down the style, its travel through the micropyle and its entrance into the nucellus.
- The two sperms are released into the embryo sac. One sperm nucleus unites with the egg nucleus to give rise to a 2*n* zygote (embryo).
- The other sperm fuses with the secondary nucleus to form an endosperm nucleus with triploid chromosome number.
- From this double fertilization, a seed develops. Higashiyama et al. (2001) demonstrated by laser cell ablation in flowering plants that two synergid cells adjacent to the egg cell attract pollen tubes.
- Once fertilization is completed, the embryo sac no longer attracts the pollen tube and this cessation of attraction might be involved in blocking polyspermy.

REFERENCES

Singh, R. J., and G. Röbbelen. 1975. Comparison of somatic Giemsa banding pattern in several species of rye. Z. *Pflanzenzüchtg,* 75: 270–285.

Singh, R. J., G. Röbbelen, and M. Okamoto. 1976. Somatic association at interphase studied by Giemsa banding technique. *Chromosoma (Berlin)*, 56: 265–273.

Smith, M. U., and A. C. H. Kindfield. 1999. Teaching cell division Basics & recommendations. *Am. Biol. Teach.* 61: 366–371.

8 Mode of Reproduction in Plants

8.1 SEXUAL REPRODUCTION

- The mode of reproduction in plants may be sexual (amphimixis), asexual (apomixis or agamospermy), or by specialized vegetative structures (vegetative reproduction).
- Sexual reproduction is a complex biological activities that facilitates genetic diversity and speciation.
- In flowering higher plants, the reproductive organs are in the flower. Meiosis and fertilization are two essential processes in the sexual cycle of higher plants.
- Sexual reproduction consists of two generations—sporophytic and gametophytic.
- The sporophytic generation begins when an egg nucleus unites with a sperm nucleus producing an embryo and the second sperm nucleus fuses with the polar nuclei (secondary nucleus) producing triploid endosperm. This continues with the development of seed, seedling, mature plant, and flowers (Figure 8.1).
- Sporophytic or somatic tissues contain diploid ($2n$) chromosome number. The flower contains spore-forming organs called anthers and ovaries through meiosis.
- Anthers and the ovaries produce, respectively, haploid (n) microspores and megaspores. Thus, alternation of sporophytic and gametophytic generations is a rule in the sexual reproduction (Figure 8.1).
- Sexual reproduction involves pollination, germination of pollen grains on the stigma, growth of pollen tubes down the style, the entrance of pollen tube to the ovule through the micropylar opening, the discharge of two sperm nuclei into the embryo sac, and fusion of a sperm nucleus with the egg nucleus producing an embryo and conjugation of a second sperm nucleus with the two polar nuclei (secondary nucleus) forming a triploid endosperm.
- Thus, seeds develop as the result of double fertilization and the majority of plants reproduce sexually (Figure 8.1).

8.2 ASEXUAL REPRODUCTION

- The disruption of the sexual process may lead to apomixis.
- Apomictic plants of some species produce seed directly from chromosomally unreduced megaspore mother cells or from somatic cells of the nucellus or ovule without fertilization (Figure 8.2).

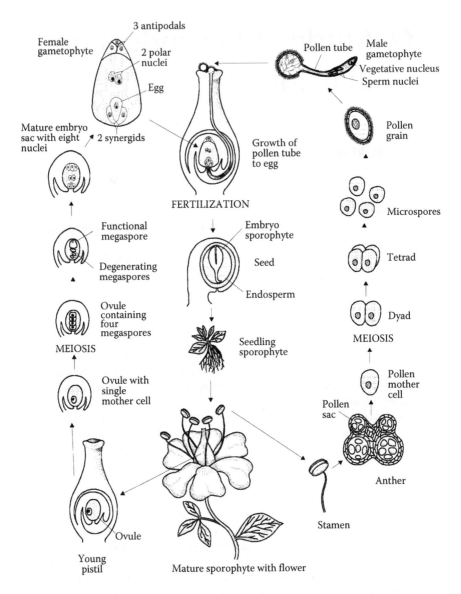

FIGURE 8.1 A diagrammatic sketch of the sexual life cycle (amphimixis) in plants. (Redrawn from Nelson et al. *Fundamental Concept of Biology,* 2nd ed., John Wiley, NY. p. 189, 1970. With permission.)

- In addition, several plants propagate by specialized vegetative structures like bulbs, corms, cuttings, runners, rhizomes, tubers, and by grafting.
- The majority of wild polyploid perennial relatives of cultivated species harbor the apomixis.
- Only a small proportion of cultivated crops in natural populations are apomictic.

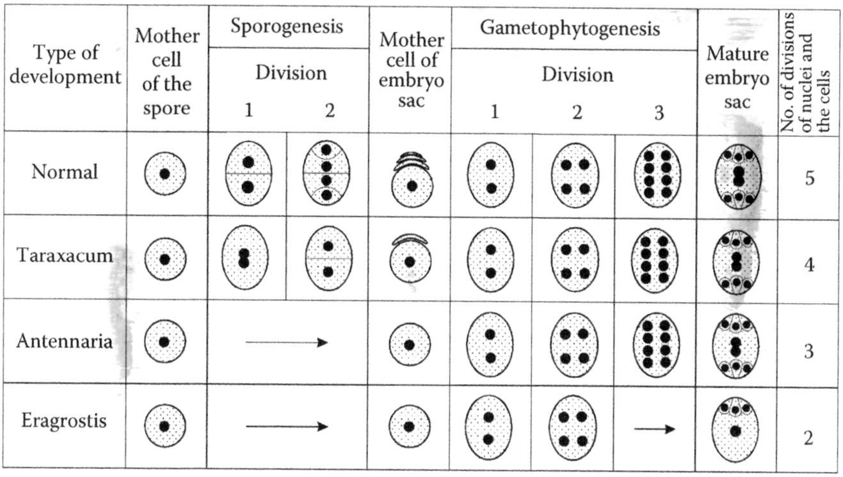

Type of development	Mother cell of the spore	Sporogenesis		Mother cell of embryo sac	Gametophytogenesis			Mature embryo sac	No. of divisions of nuclei and the cells
		Division			Division				
		1	2		1	2	3		
Normal	(•)	(:)	(⫶)	(≋•)	(:)	(⫶)	(⫶⫶)	(⬤)	5
Taraxacum	(•)	(⦙)	(:)	(⊘)	(:)	(⫶)	(⫶⫶)	(⬤)	4
Antennaria	(•)	⟶		(⊙)	(:)	(⫶)	(⫶⫶)	(⬤)	3
Eragrostis	(•)	⟶		(⊙)	(:)	(⫶)	⟶	(⬤)	2

FIGURE 8.2 Diagrammatic sketch of megagametogenesis in amphimixis and various apomictic plants.

- Apomixis is very rare in the diploid species and over 90% of the apomictics are usually polyploids.
- Apomixis predominates in cross-fertilizing plants and has not developed in self-fertilizing plants.
- Wild relatives of wheat, maize, and pearl millet and many tropical grass genera carry genes for apomixis.

8.2.1 TYPES OF APOMIXIS

8.2.1.1 Gametophytic Apomixis
- It includes diplospory and apospory.
- Diplospory is divided into *Taraxacum* (Figure 8.2) and *Antennaria* (Figure 8.2).
- *Taraxacum* type apomixis, nucleus divides but does not move to the poles and second sporogenesis division occurs producing two cells (Figure 8.2). This is found in the common dandelion (*Taraxacum officinale* F. H. Wigg; Figure 8.3).
- In *Antennaria* type, sporogenesis is absent and nucleus enters into gametophytogenesis (Figure 8.2).
- In *Eragrostis* type, division of the mother cell of the spore in sprogenesis is absent and third division of gametophytogenesis is also absent producing an embryo sac with antipodal cells and a secondary nucleus. Thus, the number of divisions of nuclei and the cells is two times (Figure 8.2).
- Apospory includes *Hieracium* and *Panicum*.
  - This apomixis maintains the alternation of generation.
  - Meiosis and fertilization are circumvented in both forms and an unreduced egg cell develops asexually by parthenogenesis.

(a)

(b)

FIGURE 8.7 (a) The tassel of a maize plant at pollen shedding stage with large protruding and hanging anthers. Female (ears) organ is on the same plant; silk includes style and stigma corresponding to each kernel are out from the husk leaf; (b) a tassel showing hanging anthers in the field of maize.

- *Self-incompatible plants*: Both male and female spores are produced at the same time and are functional but fail to fertilize and produce seeds; entirely cross-pollinated.
- *Unisexual*: Plants having either functionally male or functionally female flowers. This is also called incomplete or imperfect flowers.

8.4 APOMIXIS IN CROP IMPROVEMENT

- Apomixis helps fix heterosis in a desired heterozygous gene combinations with valuable agronomic traits and resistance to pests and pathogens.
- Apomixis use in crop improvement is limited to turf and forage grasses.

8.5 VEGETATIVE REPRODUCTION

- Vegetative reproduction is an asexual form of reproduction in higher plants by which new identical individuals are generated from a single parent without sexual reproduction.
- The offspring of asexual propagation is known as a clone.
- It is characterized by mitosis which occurs in the shoot and root apex, cambium, intercalary zones, callus tissues, and adventitious buds.

8.5.1 VEGETATIVE REPRODUCTION BY SPECIALIZED VEGETATIVE ORGANS

8.5.1.1 Bulb
- Short basal, underground stem surrounded with thick, fleshy leaves.
- Common in the family of Allioideae (Allium-lily, tulip, onion, garlic). Figure 8.8 shows a bunch of shallots lying on the kitchen floor being examined for cooking by Kingston (my *grandson*). Figure 8.9 shows a longitudinal section of an onion bulb showing fleshy leaves.

8.5.1.2 Corm
- It is also known as bulbotuber; a corm is short, upright, with a hard or fleshy bulb-like stem usually covered with papery, thin dry leaves; it does not contain fleshy leaves.
- It is common in families Alismataceae, Araceae, Aspergaceae, Asteraceae, Colchicaceae, Cyperaceae, Iridaceae, and Musaceae. Gladiolus (Figure 8.10; https://en.wikipedia.org/wiki/Corm#/media/File:Corm.jpg), banana, water chestnut (Figure 8.11 [http://www.chesapeakebay.net/images/field_guide/Water_Chestnut_page_image.jpg], and taro [*Colocasia esculanta* (L.)] Schott; Figure 8.12).

8.5.1.3 Runners (Stolons)
- Runners are a horizontal above-ground stem that usually produce plants by rooting at nodes (Figure 8.13).

FIGURE 8.8 Several bunches of shallot (bulb) being on the kitchen floor being examined by Kingston (my grandson) for quality. Shallots in bags produce nice roots used for cytological studies (Chapter 2).

FIGURE 8.9 Horizontal cross-section of onion.

FIGURE 8.10 Gladiolus is an example of corm.

FIGURE 8.11 Water chestnut growing in a pond; corm.

FIGURE 8.12 Germinating taro rhizomes.

FIGURE 8.13 Brachiaria grass growing in the field showing adventitious roots at each node.

- Strawberry reproduces vegetatively via stolons.
- Adventitious roots at the nodes form roots and new plants from the node.
- Plants with stolons are called stoloniferous.
- Stolons develop from base of the plants.
- In strawberries, the base is above the soil surface. The stolons remain underground and form shoots that rise to the surface at the ends or from the nodes. The nodes of the stolons produce roots (https://en.wikipedia.org/w/index.php?curid=2729259).

8.5.1.4 Rhizomes
- A rhizome is a horizontal, prostrate, or underground stem that contains nodes and internodes of various lengths and readily produces adventitious roots.
- Rhizome species are easily propagated by cutting the rhizomes into small pieces that contain a vegetative bud.
- Propagation through rhizomes includes hops, asparagus, turmeric (Figure 8.14), ginger (Figure 8.15), cannas, orchids, Johnson grass, and Bermuda grass.

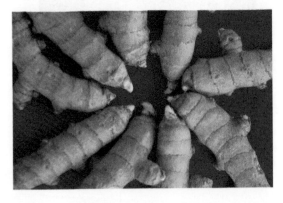

FIGURE 8.14 Growing roots of rhizome turmeric. (Courtesy of Nirmal Babu.)

(a)

(b)

FIGURE 8.15 Rhizomes of ginger grown in a pot (a) in the greenhouse showing actively growing buds (b).

8.5.1.5 Tubers

- A tuber is fleshy portion of a rhizome underground storage stem.
- Potato (Solanaceae) (Figures 8.16 and 8.17), sweet potato (Convolvulaceae; Figure 8.18), cassava (Euphorbiaceae), yam (Dioscoreaceae), and ginseng (Figure 8.19) are the best examples of a species with tuber although these tuber crops belong to different families.

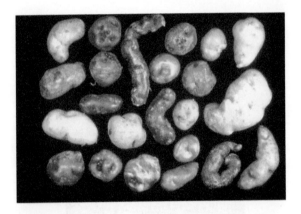

FIGURE 8.16 Various sizes, shapes, and colors of potato tubers.

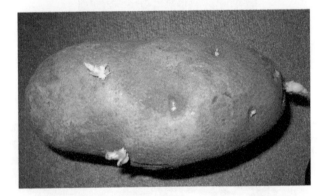

FIGURE 8.17 Germinating potato.

FIGURE 8.18 Freshly harvested sweet potato from the farm of Professor Chung's brother in Jinju, Korea. This photo was snapped during my 2010 visit.

FIGURE 8.19 Fleshy tuber of ginseng; photo was taken in a market of Jinju, Korea.

- Figure 8.17 shows a potato tuber with several eyes but only few are showing sprouting shoots.
- Other examples include dahlia and begonia; the primary tap root develops into an enlarged tuberous root that can be propagated by dividing into several portions containing a bud.

8.5.2 Vegetative Reproduction by Adventitious Roots and Shoots

- Reproduction of an entire plant from a buried branch or stem is called layering.
- Once new roots and shoots emerge, plantlets are separated from the mother plants.
- Cutting is one of the most important methods of vegetative reproduction.
- Small pieces of stems (cuttings); Figure 8.20 are used by the housewife, horticulturist, and nurseryman for multiplying and reproducing ornamental crops. Cutting is routinely used in sugarcane cultivation (Figure 8.21). Cuttings with 2–3 buds are directly planted in the field during the month of March; I have seen these in my home town (Sirihara).

8.5.3 Vegetative Reproduction by Grafting

- Reproduction of an entire plant by union of small actively growing shoot (scion) grafted onto root stock that is resistant to pathogens and pests in an invaluable tool in plant propagation.
- Grafting is quite common for a large number of domestic fruit crops to produce disease-free crops.
- Grafting in mango, apple, and grape is routinely used.

FIGURE 8.20 Rooting of cutting in a derived line of wide hybrid between *G. max* cv. Dwight x *G. tomentella*, PI 441001. This method helps in maintaining the sterile hybrids of intersubgeneric hybrids.

FIGURE 8.21 Propagation of sugarcane by cutting (http://www.canarius. com/en/search?controller=search&orderby=position&orderway=desc&search_ query=sugar+cane&submit_search=).

8.5.4   VEGETATIVE REPRODUCTION BY TISSUE CULTURE

- Plant propagation through cell and tissue cultures is also known as micropropagation.
- It involves regeneration of plants aseptically from cells (including protoplasts) and tissues (immature embryos [Figure 8.22], leaves, roots, and stems) in artificial cultures.
- A single protoplast can regenerate an entire new plant.
- This method generates a large number of plantlets via embryogenesis and organogenesis from a small piece of stock plant.
- Micropropagation has proven efficient for orchid propagation where the natural propagation rate is very slow.

8.5.5   ADVANTAGES OF VEGETATIVE REPRODUCTION

- Vegetative propagation has numerous advantages such as efficient production of potato and sugarcane.
- Exploitation of heterosis
- Avoidance of dormancy and juvenile period (grafting on older stocks allows new wood of seedling to produce fruit sooner than if it remains on its own root stock.)
- Maintenance of sterile or lethal genotypes
- Facilitation of physiological and genetic studies
- Increase plants of unique genotypes
- Despite many advantages, vegetative reproduction has a serious problem. All vegetative propagated plants from the same source are genetically uniform.

FIGURE 8.22 Multiple shoot in an intersubgeneric hybrids of *G. max* cv. Dwight x *G. tomentella*, PI 441001.

This means that genetic vulnerability exists. If the genotype is susceptible to a pest or pathogen, or if a new pest or a pathogen develops that can infect the genotype, then all plants of the clone will be susceptible. If a disease strikes the members of a clone, production of a disease-free seed or shoot is extremely difficult.

8.6 METHODOLOGIES FOR PRODUCING AMPHIDIPLOIDY

8.6.1 COLCHICINE

Of all the above means of obtaining tetraploids (polyploidy), colchicine has been found to be the most effective chemical to induce polyploidy in a large number of plant and animal species (Blakeslee and Avery 1937; Blakeslee 1939; Eigsti and Dustin 1955; Burnham 1962). Colchicine is an alkaloid and is a highly poisonous (carcinogenic) chemical that should not be absorbed by the skin (Blakeslee and Avery 1937). Colchicine is extracted from the seeds and bulbs of the wild meadow saffron or autumn-flowering crocus, (*Colchicum autumnale* L. [Eigsti and Dustin 1955]).

- Colchicine acts by inhibiting spindle formation and preventing anaphase. Chromosomes stay at the equatorial plate but split longitudinally and divided chromosomes remain in a single restitution nucleus doubling the chromosome number. This process will be continued as long as the drug is present.
- Colchicine is applied to growing organs of plants, to rapidly germinating seeds, to growing shoots, and buds and axillary nodes. Colchicine is water-soluble.
- Seed treatment: Colchicine concentrations and treatment durations depend upon the crop species. Blakeslee and Avery (1937) treated rapidly germinating *Datura* seeds with colchicine concentrations ranging from 0.003125% to 1.6% for 10 days. Concentrations up to 0.1% were not effective but tetraploids were obtained in the higher concentrations (0.2% to 1.6%).
- Seed treatment with 0.25% aqueous colchicine was found to be the most effective in producing autotetraploids in chick pea (*Cicer arietinum* L., 2n = 16), but seedling treatment failed (Pundir et al. 1983). Colchicine-treated seeds and seedlings should be washed thoroughly in running tap water before planting.
- Seedling treatment: Actively growing shoots of young seedlings may be immersed in aqueous colchicine solution or colchicine may be applied to shoots by absorbent cotton or swab. Immersion of roots should be avoided. Sears (1941) obtained amphidiploid sectors from sterile intergeneric hybrids of Triticinae by use of colchicine. The crowns and bases of plants were wrapped with absorbent cotton and the cotton was soaked with 0.5% aqueous solution of colchicine. The plants were transferred to a high humidity chamber to maintain humidity.
- Treated plants can be covered with clear plastic bags. Cotton was kept wet for 2–5 days by applying colchicine twice a day. Cotton was removed and plants were transferred to the greenhouse for further growth. Tetraploid sectors

were obtained in 39 of the 60 surviving plants, representing 17 different hybrids.

- Stebbins (1949) isolated autotetraploids from 20 different species of the family Poaceae (Gramineae) by treating shoots with 0.1%–0.2% aqueous solution of colchicine for 8–24 h. Roots were washed thoroughly before planting. However, root feeding of colchicine to *Glycine* species hybrids failed to produce amphiploids but shoot treatment was successful.

- Schank and Knowles (1961) recorded that the most successful colchicine treatment to induce tetraploidy in safflower was a 0.1% aqueous solution applied to a cotton swab wedged between cotyledons of 3-day-old seedlings. The shoot treatment was less successful.

- Thiebaut et al. (1979) recommended treating barley haploid seedlings at the three-leaf stage using a solution containing 0.1% colchicine, 2% DMSO, 0.3 mL/L (10 drops) of Twin 20, and 10 mg/L GA3 under 25°C–32°C. Seedlings were treated for 5 h.

8.6.2 Nitrous Oxide

- It has been demonstrated that nitrous oxide is a chromosome doubling agent. Compared to colchicine it induces a relatively higher frequency of chromosome-doubled plants, causes less lethality, and the gas is relatively harmless. Chromosome doubling by nitrous oxide treatment is very effective for plants still in tissue culture but is unsuitable for large plants (Hansen et al. 1988).

- Taylor et al. (1976) excised heads of diploid red clover containing 2 cm stems 24 h after crossing and placed them in vials containing a 2% aqueous sucrose solution. Vials with heads were placed in a gas-tight chamber and nitrous oxide was maintained to 6 bars atmospheric pressure (approximately 90 psi).

- After 24 h, the heads were removed to a dark incubator at 20°C for a seed maturation period of 3 weeks. A total of 226 plants were treated with nitrous oxide and 160 plants (71%) were identified as putative tetraploid ($2n = 4x = 28$) based on pollen size. Of 136 plants examined cytologically, 119 plants were tetraploids, 3 plants were diploid, 2 plants died, and 12 plants were aneutetraploid ($2n = 26, 27,$ or 29). No plants with chromosomal chimaerism were found.

- Hansen et al. (1988) compared the effect of nitrous oxide and colchicine on chromosome doubling of anther culture-derived young seedlings of wheat still in culture. Two colchicine treatments (0.005% and 0.01% for 24 h) were compared with two nitrous oxide treatments (24 and 48 h at 6 atm).

- Both nitrous oxide treatments were as effective as 0.01% colchicine; however, colchicine treatment killed a significant proportion of the treated plants, while nitrous oxide treatment was nontoxic. The low concentration of colchicine (0.005%) showed lower chromosome doubling efficiency.

- CC BY-SA 3.0, https://commons.wikimedia.org/w/index.php?curid=600604

8.7 BARRIERS TO CROSSABILITY IN PLANTS

- Geographical separation
- Timing of flowering due to day length
- Species flowering at different length
- Varietal differences preventing hybridization
- Self- and cross-incompatibility
- Separation of sexes
- Failure of pollen tube to germinate
- Bursting of pollen tubes
- Pollen tube growth too slow to reach ovary
- Pollen tube reaching to ovary but fertilization does not take place
- Fertilization occurs but embryo development is arrested
- Embryo develops for few days but viable seeds are not formed
- Disharmony in parents produces inviable and sterile plants

REFERENCES

Aron, Y., H. Czosnek, and S. Gazit. 1998. Polyembryony in mango (*Mangifera indica* L.) is controlled by a single dominant gene. *Hort. Sci.* 33: 1241–1242.

Blakeslee, A. F. 1939. The present and potential service of chemistry to plant breeding. *Amer. J. Bot.* 26: 163–172.

Blakeslee, A. F., and A. G. Avery. 1937. Methods of inducing doubling of chromosomes in plants by treatment with colchicine. *J. Hered.* 28: 393–411.

Burnham, C. R. 1962. *Discussion in cytogenetics*. Burgess, Minneapolis, MN.

Eigsti and Dustin, 1955. *Colchicine in Agriculture, Medicine, Biology and Chemistry*. Iowa State Univ. Press, Ames, IA.

Hansen, F. L., S. B. Anderson, I. K. Due, and A. Olesen. 1988. Nitrous oxide as a possible alternative agent for chromosome doubling of wheat haploids. *Plant Sci.* 54: 219–222.

Koltunow, A. M., T. Hidaka, and S. P. Robinson. 1996. Accumulation of seed storage proteins in seeds and in embryos cultured in vitro. *Plant Physiol.* 110: 599–609.

Naumova, T. N. 1993. *Apomixis in Aangiosperms. Nucellar and Integumentary Embryony*. CRC Press, Inc., Boca Ratan, FL.

Pundir, R. P. S., N. K. Rao, and L. J. G. van der Maesen. 1983. Induced autotetraploidy in chickpea (*Cicer arietinum* L.). *Theor. Appl. Genet.* 65: 119–122.

Schank, S. C., and P. F. Knowles. 1961. Colchicine induced polyploids of *carthamus tinctorius* L. *Crop Sci.* 1: 342–345.

Sears, E. R. 1941. Amphidiploids in the seven-chromosome *Triticinae*. Mo. Agric. Exp. Stn. Res. Bull. 336: 1–46.

Stebbins, G. L. Jr. 1949. The evolutionary significance of natural and artificial polyploids in the family Gramineae. *Proc. 8th Int. Congr. Genet.* 33: 461–485.

Taylor, N. L., M. K. Anderson, K. H. Quesenberry, and L. Watson. 1976. Doubling the chromosome number of *Trifolium* species using nitrous oxide. *Crop Sci.* 16: 516–518.

Thiebaut, J., K. J. Kasha, and A. Tsai. 1979. Influence of plant development stage, temperature, and plant hormones on chromosome doubling of barley haploids using colchicine. *Can. J. Bot.* 57: 480–483.

9 Karyotype Analysis

9.1 INTRODUCTION

- Cytological techniques determine the chromosome constitution of an organism and facilitate recognition of the individual chromosomes.
- Three terms, namely karyotype, karyogram, and idiogram, are often referred to in the identification of chromosomes.
- Karyotype is the number, size, and morphology of the chromosome set of a cell in an individual or species (Battaglia 1994).
- Karyogram is the physical measurement of the chromosomes from a photomicrograph where chromosomes are arranged in descending order (longest to shortest). An idiogram represents a diagrammatic sketch (interpretive drawing) of the karyogram (Figure 9.1).
- The classification of chromosomes is based on physical characteristics, such as size of chromosomes, features of telomere, position of kinetochore, secondary constriction, size and position of heterochromatic knobs, and relative length of chromosomes (Figure 9.2).

9.2 NOMENCLATURE OF CHROMOSOMES

- The kinetochore (centromere) position is a very useful landmark for the morphological identification and nomenclature of chromosomes.
- Monocentric chromosomes contain one centromere and nomenclatured chromosomes are based on the position of the centromere in the chromosomes:
  - *Median centromere (isobrachial chromosomes)*: Centromere is situated in the middle of the chromosome resulting in an arm ratio of 1:1.
  - *Submedian centromere (heterobrachial chromosome)*: Centromere is located near the middle of the chromosome resulting in an arm ratio of more than 1:1 but less than 1:3 (from 1:1 to 1:2.9).
  - *Subterminal centromere (hyperbrachial chromosome)*: Centromere is near one extremity of the chromosome resulting in ratio of 1:3 or more.
  - *Terminal centromere (monobrachial chromosome)*: Centromere is situated at one extremity of the chromosome resulting in an arm ratio of 0:1.
- A satellite as a part of the chromosome distal to a nucleolar constriction and it is universally accepted that for each satellite there is one nucleolus. It is an established fact that a satellite has a spheroidal body and diameter that is either the same or smaller than the diameter of the chromosomes, situated at one extremity, connected to the chromosome body by a thin thread (Figure 9.3).

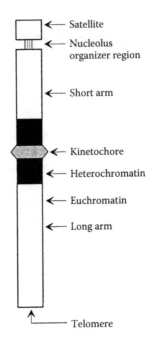

FIGURE 9.1   An ideogram (diagrammatic sketch) of a metaphase chromosome.

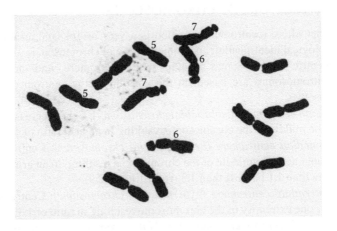

FIGURE 9.2   Somatic mitotic metaphase chromosomes of barley after acetocarmine staining showing $2n = 14$ chromosomes.

- The position of the satellite may be terminal (a satellite between its nucleolar constriction and one extremity) or intercalary (a satellite between two
  nuclear constrictions). A satellite chromosome is designated based on size
  and location (Figure 9.3):
  - *Microsatellite*: A spheroidal satellite of small size, that is, having a
    diameter equal or less than one half the chromosomal diameter.

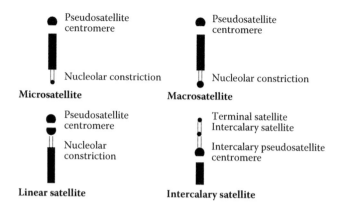

FIGURE 9.3 Diagrammatic sketch of the variability of chromosome morphology based on location of secondary nucleolar constriction. (From Battaglia, E., *Caryologia,* 8, 179–187, 1955. With permission.)

- *Macrosatellite*: A spheroidal satellite of large size, that is, having diameter greater than one half the chromosomal diameter.
- *Linear satellite*: A satellite having the shape of a long chromosomal segment.

9.3 KARYOTYPE ANALYSIS BY MITOTIC METAPHASE CHROMOSOMES

- Karyotype analysis has played an important role in the identification and designation of chromosomes in many plant species. Barley (*Hordeum vulgare* L.) is cited here as an example. Barley is a basic diploid, contains $2n = 2 = 14$ chromosomes
- Chromosome 5 is the smallest chromosome and chromosomes 6 and 7 are the nucleolus organizer chromosomes and are morphologically distinct. Based on conventional staining techniques, chromosomes 1–4 are difficult to distinguish (Figure 9.2).
- The application of the Giemsa C- (Figure 9.4) and N-banding techniques helped to identify all the barley chromosomes while it was not possible by conventional staining techniques.
- The salient features of the seven barley chromosomes based on conventional and Giemsa C- and N-banding techniques and homoeologous (in parenthesis) groups (Costa et al. 2001) with wheat are described below.

9.3.1 CHROMOSOME 1 (7H)

This is the third longest chromosome and is a metacentric (Figure 9.2). Since both arms are almost equal in measurement, their designation in karyogram and idiogram depended on the Giemsa N-banding pattern (Figure 9.5).

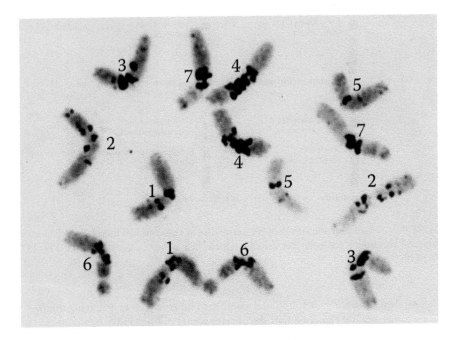

FIGURE 9.4 Somatic mitotic metaphase chromosomes of barley after Giemsa C-banding. (From Singh, R. J. and Tsuchiya, T., *Z. Pflanzenüchtg,* 86, 336–340, 1981b. With permission.)

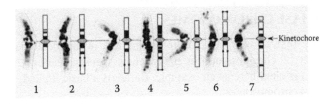

FIGURE 9.5 Karyogram and idiogram of Giemsa N-banded chromosomes of barley. (Redrawn from Singh, R. J., and Tsuchiya, T., *J. Hered.,* 73, 227–229, 1982b. With permission.)

9.3.2 CHROMOSOME 2 (2H)

- This is the longest chromosome among the five non-satellite chromosomes of the barley complement.
- It carries its kinetochore at the median (l/s = 1.26) region.

9.3.3 CHROMOSOME 3 (3H)

- It is a median (arm ratio = 1.09) chromosome.
- It showed a dark centromeric band. The band on the short arm appeared as a large block at metaphase. The long arm had a dark interstitial band (close to the kinetochore) and a faint dot on each chromatid in the middle of the long arm (Figure 9.5).

9.3.4 CHROMOSOME 4 (4H)

- This chromosome contains its kinetochore at the median region (l/s = 1.21), and was correctly identified in all the studies.
- Conventional staining techniques do not distinguish chromosome 4 from chromosomes 1, 2, and 3.
- However, based on Giemsa C- and N-banding techniques, chromosome 4 was easily distinguished from the rest of the chromosomes because it is the most heavily banded in the barley complement; about 48% of the chromosome is heterochromatic.
- Sometimes, it is difficult to locate the centromere position in condensed Giemsa-banded metaphase chromosomes (Figure 9.5). However, the appearance of a diamond-shaped centromere position and the use of aceto-carmine stained Giesma N-banding technique facilitated the precise localization of the kinetochore (Figure 9.5).

9.3.5 CHROMOSOME 5 (1H)

- This chromosome is the shortest among the five nanosatellite chromosomes of barley and has an arm ratio (1.42) similar to chromosome 3.
- It has a centromeric band, and an intercalary band on the long arm, and a band on the short arm that is darker than those of long arm (Figure 9.5).

9.3.6 CHROMOSOME 6 (6H)

- This chromosome has a larger satellite than chromosome 7 and has an arm ratio of 1.66 (without the satellite).
- Chromosome 6 showed a dark centromeric band in both arms, a faint intercalary band on the long arm and a faint dot on each chromatid on the telomere of the satellite (Figure 9.5).

9.3.7 CHROMOSOME 7 (5H)

- This chromosome has the longest long arm in the barley karyotype and carries a submedian kinetochore.
- It showed an equally dense centromeric band at the distal portion of the long arm and a faint intercalary band was also observed in the short arm (Figure 9.5).

9.4 KARYOTYPE ANALYSIS BY PACHYTENE CHROMOSOMES

- The classical examples for conducting karyotype analysis based on pachyenema chromosomes are in maize, tomato, *Brassica,* and rice and soybean.
- Pachynema chromosomes are differentiated by heterochromatin and euchromatin shown in soybean (Figure 9.6).

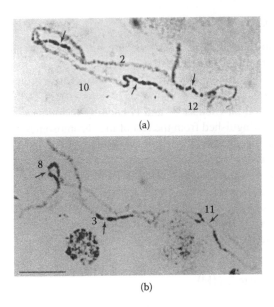

(a)

(b)

FIGURE 9.6 Pachytene chromosomes of soybean; (a) three soybean chromosomes (2, 10, and 12); (b) chromosome 3, 8, and 11. Arrow shows kinetochore flanked by heterochromatin.

- Based on an examination of 20 pachynema chromosomes of soybean, euchromatin and heterochromatin distribution, showing contrasting differences in chromosome length 20 chromosomes were cleverly identified (Figure 9.7).
- Sometimes, pachynema chromosomes are used when somatic chromosomes do not show distinguishing landmarks. The classical example is soybean chromosomes; all chromosomes are metacentric and chromosome size is not more than 2.84 μm.

9.5   KARYOTYPE ANALYSIS BY FLOW CYTOMETRY

- Flow cytogenetics may be defined as the use of flow cytometry to sort and analyze individual chromosomes and physical mapping of genes of economic importance using FISH technology. However, this technique has not been universally successful in plants because of the lack of quality chromosomes and an inability to resolve a single chromosome on flow karyotype.
- Lucretti et al. (1993) invented a procedure to sort only the metacentric chromosome of *Vicia faba* by flow cytometry. They located rDNA locus by FISH. Preparation of high-quality chromosome suspensions is a prerequisite for successful chromosome sorting and karyotyping. They listed several factors for poor quality chromosomes in suspensions, such as the splitting of metaphase chromosomes into chromatids, chromosome breakage, chromosome clumping, presence of interphase nuclei, and presence of cellular and chromosomal debris.

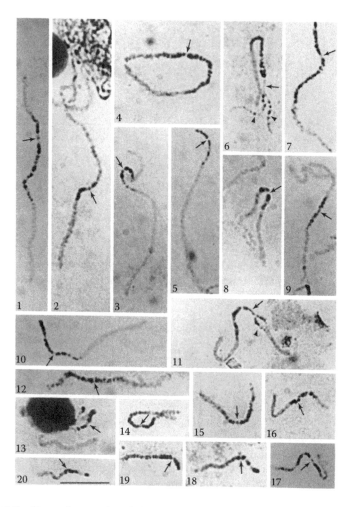

FIGURE 9.7 Photomicrographs of the pachytene chromosome complement of *Glycine max* x *G. soja* F₁ hybrid. Each figure shows a different chromosome. For example, 1, chromosome 1–20, chromosome 20. Arrows indicate centromere location. (From Singh, R.J., and Hymowitz, T., *Theor. Appl. Genet.*, 76, 705–711, 1988. With permission.)

- Flow cytometry may be an extremely valuable tool if we can distinguish aneuploid from diploid plants based on relative surplus or deficit of DNA content. Taking this view into consideration, Samoylova et al. (1996) examined a relative surplus of DNA content in *Arabidopsis* primary and telotrisomics in interphase nuclei measured by flow cytometer to distinguish diploid (wild-type) from trisomic plants. They measured differences in nuclear fluorescence intensity between diploid and trisomics.
- The relative surplus of genomic DNA recorded by primary and telotisomics was attributed to the extra chromosome (Figure 9.8). However, flow karyotype contradicts cytological observation. Cytologically chromosomes 5 and 3 are

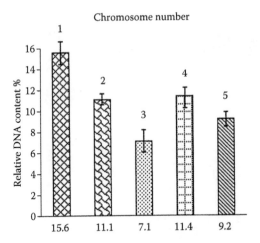

FIGURE 9.8 Diagrammatic representation of the flow karyotype of *Arabidopsis thaliana* interphase chromosomes. (From Samoylova, T. I., et al., *Plant J.*, 10, 949–954, 1996. With permission.)

larger than chromosomes 4 and 2 (the smallest) but relative DNA content (%) was found lesser for chromosomes 5 and 3 than 4 and 2 (Figure 9.8). At this stage we may conclude that flow karyotyping for plants needs perfecting.

9.6 KARYOTYPE ANALYSIS BY IMAGE ANALYSIS

- The beginning of chromosome image analysis goes back to the early 1980s when computer systems were at the cradle stage to handle the huge digital data of images.
- There were only a few expensive image analyzing systems available and imaging techniques suitable for plant chromosomes analysis were under development.
- In 1985, the first comprehensive chromosome image analyzing system (CHIAS) with software fulfilling the basic requirements of cytologists and cytogeneticists was developed (Fukui 1986). Then further development of imaging methods, such as quantifying chromosome morphology and its band patterns in barley (Fukui and Kakeda 1990), quantifying uneven condensation patterns appearing at the prometaphase chromosomes in rice (Fukui and Iijima 1991), and simulating human vision for identifying and quantifying chromosome band patterns in *Crepis* (Fukui and Kamisugi 1995) followed. Imaging method for quantification of pachytene chromomeres was soon released.
- No personal computers with enough imaging capability allow image analysis in every cytology and cytogenetics laboratory. The basic points that should be borne in mind for the application of imaging methods are as follows:
  a. Importance of the quality of chromosome images. No imaging method can create new information that is not originally included in the chromosome images.

b. The information of the images is basically reduced by each application of image manipulation. Imaging methods present essence of the chromosome image of the original image as visible and thus perceptible way.

c. Imaging methods can present the image information by numerical data.

The standard and basic procedures for image analyses and the manuals (Kato et al., 1997) can be obtained either by written form or via the Internet (http://mail. bio.eng.osaka-u.ac.jp/cell/).

REFERENCES

Battaglia, E. 1994. Nucleosome and nucleotype: A terminological criticism. *Caryologia* 47: 193–197.

Costa, J. M., A. Corey, P. M. Hayes, C. Jobet, A. kleinhofs, A. Kopisch-Obusch, S. F. Kramer, D. Kudrna, M. Li, O. Riera-Lizarazu, et al. 2001. Molecular mapping of the Oregon Wolfe barleys: A phenotypically polymorphic doubled-haploid population. *Theor. Appl. Genet.* 103: 415–424.

Fukui, K. 1986. Standardization of karyotyping plant chromosomes by a newly developed chromosome image analyzing system (CHIAS). *Theor Appl. Genet.* 72: 27–32.

Fukui, K., and K. Iijima. 1991. Somatic chromosome map of rice by imaging method. *Theor. Appl. Genet.* 81: 589–596.

Fukui, K., and K. Kakeda. 1990. Quantitative karyotyping of barley chromosomes by image analysis methods. *Genome.* 33: 450–458.

Fukui, K., and Y. Kamisugi. 1995. Mapping of C-banded Crepis chromosomes by image methods. *Chromosome Res.* 3: 79–86.

Lucretti, S., J. Doležel, I. Schubert, and J. Fuchs. 1993. Flow karyotyping and sorting of *Vicia faba* chromosomes. *Theor. Appl. Genet.* 85: 665–672.

Samoylova, T. I., A. Meister, and S. Miséra. 1996. The flow karyotype of *Arabidopsis thaliana* interphase chromosomes. *Plant J.* 10: 949–954.

10 Classical Methods for Associating Genes with the Chromosomes

10.1 INTRODUCTION

The science of genetics relates to heredity and variation/continuity of life. Cell division, mitosis and meiosis, and precise DNA replication predicts the inheritance of a particular trait from parents to the progenies. A clear understanding of the mode of inheritance is revealed by the science of genetics and its foundation was laid by J. G. Mendel in 1865, based on experiments on the garden pea (*Pisum sativum* L.). The selection of garden pea was a wise choice because: (1) it is self-pollinated plant and pollination can be controlled, and selfing of F_1 posed no problem; (2) it is easy to cultivate this plant and requires only a single growing season; (3) it has many distinguishing sharply defined inherited traits located on different chromosomes and independently segregated (Table 10.1); (4) it has $2n = 14$ chromosomes. However, Mendel's laws were unrecognized until 1900 and this may be due to number of special reasons (Strickberger 1968). The variability among the F_2 and further hybrid generations could be traced to the original variability in the first parental cross. The factors that could be traced were followed and did not change during the period of observation, but only expressed themselves in new and different combinations among the offspring. Mendel used discontinuous traits while at that time scientists were looking for continuous variation (Galton, Darwin, and others). Mendel's approach with probability events and mathematical ratios was an unfamiliar idea to biology. Mendel was in constant correspondence with Nägeli who did not appreciate Mendel's work because he was working with *Hieracium* (hawkweed), an apomictic plant an F_1 hybrid did not segregate; all F_2 plants were identical to the maternal plant.

Mendel was fortunate not to run into the complication of linkage during his experiments. He selected seven genes of the pea which has $2n = 14$ chromosomes (seven linkage groups) and these genes are on different chromosomes. Mendel worked with two genes in chromosome 1, three genes in chromosome 4, and one gene each in chromosomes 5 and 7 (Table 10.1; Blixt 1975). Blixt (1975) assumed that out of 21 dihybrid combinations Mendel theoretically could have studied, no less than four (*a-i, v-fa, v-le, fa-le*) ought to have linkage, seeing the current genetic map of pea. It has been found that *a* and *i* in chromosome 1 are so distantly located and no linkage is normally detected and the same is true for *v, le,* and *fa* for chromosome 4. However, v and le should have shown linkage. Mendel either did not publish results or did not make the appropriate cross. Thus, Mendel did not run into the linkage complication.

TABLE 10.1

Relationship between Modern Genetic Terminology and Character Pairs Used by Mendel (Blixt 1975)

Character Pair used by Mendel	Alleles in Modern Terminology	Located in Chromosomes
Seed color: yellow-green	*I-i*	1
Seed coat and flowers: colored-white	*A-a*	1
Mature pods: smooth expanded-wrinkled	*V-v*	4
Inflorescences: from leaf axils—umbellate in top of plant	*Fa-fa*	4
Plant height: >1 m-around 0.5 m	*Le-le*	4
Unripe pods: green-yellow	*Gp-gp*	5
Mature seeds: smooth-wrinkled	*R-r*	7

Unfortunately, Mendel's work was ignored and virtually unforgotten for 34 years after its publication. But it was discovered when three scientists, DeVries, Correns, and Tschermak, independently found the same results reported by Mendel in 1866. Now, it is known as Mendelian inheritance, Mendel's laws, Mendelian factors, and others.

10.2 MONOHYBRID INHERITANCE

10.2.1 COMPLETE DOMINANCE

Mendel selected seven contrasting characters individually. He hybridized with a variety containing one different character. For example, he hybridized a round seed coat plant with a wrinkled seed shape plant. The F_1 was only one type (round seed). He allowed F_1 plants to self-pollinate. In F_2, plants produced 5474 round seeds and 1850 wrinkled. This suggests that these F_2 ratios are in close to 3:1 and can be also expressed as 3/4: ¼, or 75: 25 or 75%: 25%, shown in a checker board known as the Punnett square (Figure 10.1). Based on the results of all seven contrasting characters, Mendel concluded:

- F_1 plants expressed one type of character.
- The result was always the same regardless of which parent was hybridized.
- The trait hidden in F_1 reappeared in the F_2.

This is known as the principle or law of segregation. As the science of genetics progressed, it was established that each trait is controlled by *a gene/a* pair of genes.

10.2.2 INCOMPLETE DOMINANCE

Since the rediscovery of Mendel's law of segregation, geneticists discovered that for certain traits, heterozygous F_1 and F_2 express partial or incomplete dominance. For example, when red-flowered (*RR*) plant was hybridized with white-flowered (*rr*) plants. The F_1 plants (*Rr*) produced pink-flowered plants and after selfing F_1 plants, the F_2 plants segregated for one red (*RR*): two pink (*Rr*): one white (*rr*) (Figure 10.2).

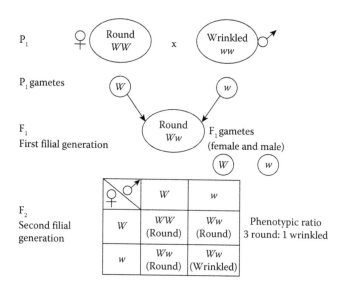

FIGURE 10.1   Monohybrid segregation showing round seed coat is complete dominant over wrinkled seed coat. F_1 plants produced seeds with round seed coat and F_2 plants segregated for three round and one wrinkled seed coat. This is Mendel's law of segregation. (Redrawn from Sinnott, E.W., Dunn, L.C., and Dobzhansky, Th., *Principles of Genetics,* McGraw-Hill Book Company, Inc. New York, 1950.)

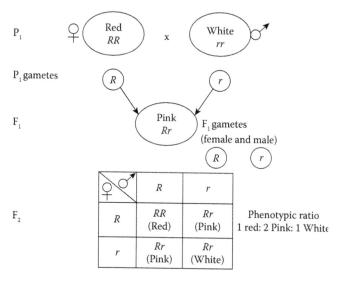

FIGURE 10.2   Sometimes, when red-flowered plants are crossed with a white-flowered plant, F_1 plants produce pink flowers and F_2 plants segregate for one red, two pink, and one white flowers. This is departure from the 3:1 ratio. However, the segregation pattern shows that red is incompletely dominant over white and heterozygous plants (Rr) produce pink flowers. (Redrawn from Sinnott, E.W., Dunn, L.C., and Dobzhansky, Th., *Principles of Genetics,* McGraw-Hill Book Company, Inc., New York, 1950.)

10.3 DIHYBRID INHERITANCE

The monohybrid hybridization (parents differing for a single gene pair) led Mendel to discover the principle of segregation. Mendel proposed the law of independent assortment by mating two parents differing in two pair of genes (dihybrid crosses). Mendel hybridized plants with yellow-round and green-wrinkled seeded plants. The F_1 plant produced yellow and round seeds. In F_2, of the 556 seeds the segregation was 315 yellow and round, 108 yellow and wrinkled, 101 green and round, and 32 green and wrinkled. This suggests that these numbers are very close to a 9:3:3:1 ratio (Figure 10.3). Thus, smooth and wrinkled, and yellow and green genes do not interfere with green and wrinkled, and two gene pairs inherit independently—*independent assortment.* We can obtain the ratio of each phenotypic combination

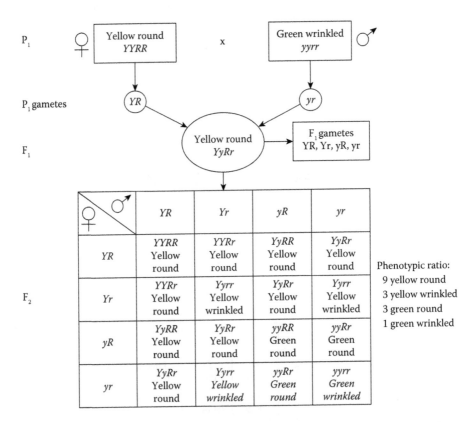

FIGURE 10.3 When two parents differ by two contrasting characters (yellow round vs. green wrinkled) F_1 plants produce yellow round seeds and in F_2 a 9:3:3:1 ratio is observed. This is law of independent segregation from Mendel. (Redrawn from Sinnott, E. W., Dunn, L.C., and Dobzhansky, Th., *Principles of Genetics,* McGraw-Hill Book Company, Inc., New York, 1950.)

by multiplying the probabilities of the individual phenotypes giving 9:3:3:1 segregation pattern (phenotypic ratio):

3/4 round × 3/4 yellow = 9/16 yellow round
3/4 round × 1/4 green = 3/16 green round
1/4 wrinkled × 3/4 yellow = 3/16 yellow wrinkled
1/4 wrinkled × 1/4 green = 1/16 green wrinkled

A question is raised about the genotypes of the segregating plants in the F_2 population. The genotype of yellow and round plants is *YYRR* and wrinkled green is *yyrr*. The plant with *YYRR* genotype will produce gametes with *YR* and *yyrr* plant will produce gametes with *yr*. The resulting F_1 plants will be yellow and round with genotype *YyRr*. The F_1 plant will produce eggs and sperms with genotype *YR*, *Yr*, *yR*, and *yr* in equal proportion (1:1:1:1). In F_2, genotypes of plants are expected as:

YYRR = 1	Yellow round	*YYRr* = 2	Yellow round
YyRR = 2	Yellow round	*YyRr* = 4	Yellow round
YYrr = 1	Yellow wrinkled	*Yyrr* = 2	Yellow wrinkled
yyRR = 1	Green and round	*yyRr* = 2	Green and round
yyrr = 1	Green wrinkled		

Mendel hybridized plants differing three genes (trihybrid cross) such as:

1. Smooth and wrinkled seed shape (*S* and *s*)
2. Yellow and green seed color (*Y* and *y*)
3. Violet and white flower (*V* and *v*)

He hybridized plants smooth, yellow, and violet (*SSYYVV*) to wrinkled, green, and white plants (*ssyyvv*). The F_1 plants expressed the phenotype of the dominant parents. The F_1 was allowed to self-pollinate and produced F_2 seeds. The F_1 hybrids produce gametes with genotype (*SYV*, *SYv*, *SyV*, *Syv*, *sYV*, *sYv*, *syV*, and *syv*). Based on gamete combinations, 8 × 8 = 64 combinations are expected. The F_2 plants produced 8 phenotypes and 27 assumed genotypes:

1. 27 smooth yellow violet = 8 genotypes
2. 9 smooth yellow white = 4 genotypes
3. 9 smooth green violet = 4 genotypes
4. 9 wrinkled yellow violet = 4 genotypes
5. 3 smooth green white = 2 genotypes
6. 3 wrinkled yellow white = 2 genotypes
7. 3 wrinkled green violet = 2 genotypes
8. 1 wrinkled green white = 1 genotype

It is interesting to note that number of gene pair differences is more than three, the number of possible combinations between them is considerably increased as summarized below:

Characteristics of segregation and independent assortment of crosses involving n pairs of alleles:

Number of Heterozygous Allelic Pairs	Number of Kinds of Gametes	Number of Phenotypes Testcross	Number of Genotypes in F_2	Number of Phenotypes in F_2[a]
1	2	2	3	2
2	4	4	9	4
3	8	8	27	8
4	16	16	81	16
n	2^n	2^n	3^n	2^n

[a] Assuming complete dominance in each allelic pair.

10.4   GENE INTERACTION AND EXPRESSION

Mendel's laws of segregation and independent assortment precisely corrects in case gene action expresses complete dominance and *lack of interference* between different genes does not occur. However, the exceptions are when traits inherit through cytoplasm, and interaction and expression of genes modify Mendelian ratios. Sometimes, dominance is not observed in some crosses, such as red (*RR*) and white (*rr*) flower plants. The F_1 plants with genotype *Rr* produce pink flower and F_2 plants segregate for red (*RR*) 1: 2 pink (*Rr*): 1 white (*rr*). Thus, genotypic ratios are phenotypic ratios (Figure 10.2).

10.4.1   LETHAL GENES

The lethal effect of a gene occurs in a particular genotype under particular set of environmental conditions. It may be dominant or recessive gene lethal as long as lethality depends upon its presence in a homozygous condition. In the house mouse, Cuénot observed that the yellow mice never breed true and concluded that yellow mice are heterozygous and hybrid. The crossing of two yellow mice produced only yellow and black and are shown as:

Yy (yellow)   x Yy (yellow)

F_1 $YYyy$

Gametes YY Yy Yy

F_2 YY lethal (mice die) (1)

                 Yy (yellow) (2) survive

                 yy (black) survive

Such mechanism has been demonstrated in several organism including human, plant, and animal. It is also known as balanced lethal.

FIGURE 10.4 White flower segregation in soybean. (a) White flower line derived from *Glycine max* cv. Dwight × *G. tomentella* PI 441001; (b) white flower accession PI 548631 (Williams; *wl* locus); (c) coleoptile of white flower is always green; (d) F_1 plants produce purple flower; (e) segregation of coleoptile in F_2 for green and purple color.

10.4.2 COMPLEMENTARY GENES

1. Duplicate recessive gene (9:7)
   - In soybeans, 07ST2-12 (Figure 10.4a) and PI 548631 (Figure 10.4b) produce white flowers and green coleoptile (Figure 10.4c).
   - The F_1 plants produced purple flowers (Figure 10.4d).
   - F_2 plants segregated for 9/16 purple (159) and 7/16 (119) green coleoptyl (Figure 10.4e). This suggests that white flower genes in parents are different.
   - The genotypes of white parents are assumed as *CCpp* and *ccPP* and the genotype of F_1 plant is *CcPp* (purple).
   - In F_2, genotypes of white flower plants are *CCpp*, *Ccpp*, *ccPP*, *ccPp*, *Ccpp*, *ccPp*, and *ccpp* (Figure 10.5).
   - The presence of *CP* is needed for the expression of purple flowers.
   - Occasionally the genes participating in such interactions are called complementary genes.

10.4.3 EPISTASIS

10.4.3.1 Dominant Epistasis (12:3:1)

When dominant allele *A* masks the expression of *B*, *A* is the epistatic gene of *B*. The gene *A* can express itself only in the presence of a *B* or *b* allele. Therefore, it is known as dominant epistasis; *B* expresses only when gene *a* is present. This phenomenon

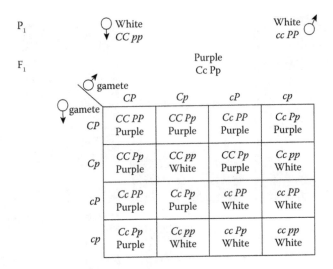

FIGURE 10.5 When white flower parents are hybridized, F_1 plants produce purple flowers and F_2 plants segregate for nine purple to seven white flowers; modification of Mendel's law of segregation. This also suggests that both white flower parents have different white flower genes. (Redrawn from Sinnott, E.W., Dunn, L.C., and Dobzhansky, Th., *Principles of Genetics*, McGraw-Hill Book Company, Inc., New York, 1950.)

of interaction between nonallelic genes is known as epistasis. In summer squashes, three common fruit colors, white, yellow, and green, are always found. In crosses between white and yellow and between white and green, white is always found to be dominant; and in crosses between yellow and green, yellow is always found to be dominant. Yellow, thus, acts as a recessive in relation to white but as a dominant in relation to green. There is evidently a gene, *W*, which is epistatic to those for yellow and green; and so long as it is present, no color is produced in the fruit, regardless of whether or not genes for color are present. In plants where the gene for white is lacking (*ww*), the fruit color will be yellow if gene *Y* is present and green if it is absent. Green-fruited plants are by *wwYY*, and white-fruited ones either by *WWYY* or by *WWyy*. This assumption is that there are two independent gene pairs, one epistatic over the other, which can be tested by crossing a homozygous white that also carries yellow, *WWYY*, with a green, *wwyy*. The F_1 plants, *WwYy*, are white-fruited. This plant will produce four kinds of gametes, *WY, Wy, wY,* and *wy.* Figure 10.5 shows expected F_2 plants; 12/16 plants will carry *W* and will be white regardless of another *Y* gene. Thus, *W* masks everything which is hypostatic to it, so that *Y*, which segregates quite independently of white, produces a visible effect only—white—in 3/16 of the plants lacking *W*. Plants (1/16) with *wwyy* gene will be green. Thus, the expected ratio is 12 white:3 yellow:1 green (Figure 10.6).

10.4.3.2 Dominant Suppression Epistasis (13:3)

When the dominant gene locus (*I*) is in homozygous (*II*) and heterozygous (*Ii*), condition and homozygous recessive alleles ii of another gene locus (*C*) produce the same phenotype. The F_2 phenotypic ratio becomes 13:3; this case is the result

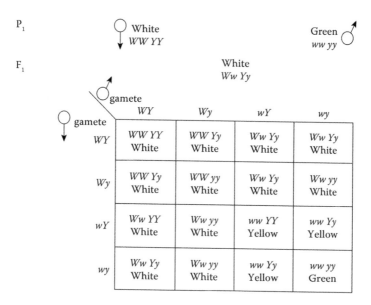

FIGURE 10.6 This checkerboard shows when white fruited is dominant over green fruited, F$_2$ plants segregate 12 white:3 yellow:1 green fruited. In this case, W gene masks everything which is hypostatic to it, Y, which segregates independently of white and expresses only when fruit lacks Y. (Redrawn from Sinnott, E.W., Dunn, L.C., and Dobzhansky, Th., *Principles of Genetics,* McGraw-Hill Book Company, Inc., New York, 1950.)

of the influence of dominant genes, which do not permit certain dominant genes to function normally. The genotype *IICC* , *IiCC*, *IIcc*, *Iicc*, and *iiccc* produce one type of phenotype (13) and genotype *iiCC* and *iiCc* will produce colored phenotype (3) (Figure 10.7).

10.4.4 COLOR REVERSION (9:3:4)

Complete dominance at both gene pairs, but one gene, when homozygous recessive, is epistatic to other. The classical example is mice coat color. When black (*CCaa*) mice are crossed with albino (*ccAA*) mice, the progenies are all agouti color (*CcPp*). When these F$_1$ agoutis are inbred, their progenies consist of 9/16 agouti, 3/16 black, and 4/16 albino (Figure 10.8).

10.4.5 COMPLETE DOMINANCE AT BOTH GENE PAIRS (15:1)

Genes with the same expression are known as duplicate genes. The classical example is the inheritance of capsule or pod form in the shepherd's purse bursa. One race has triangular capsules where as in another race, capsules are ovoid or top-shaped. When two capsule forms were crossed, F$_1$ showed triangular capsules suggesting that the triangular trait is dominant over the ovoid capsule form. In F$_2$, a 15:1 ratio is obtained; a deviation from 9:3:3:1 Mendelian ratio.

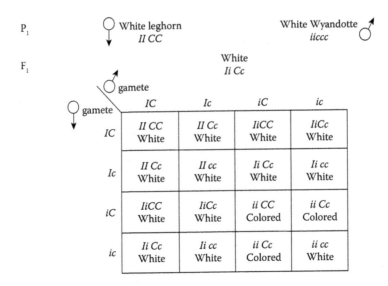

FIGURE 10.7 This Punnett square shows a 13:3 ratio which tells that if the presence of *I* gene is in recessive condition, the dominant gene *C* expresses color; thus one (*I*) is suppressing the expression of the *C* gene. (Redrawn from Sinnott, E.W., Dunn, L.C., and Dobzhansky, Th., *Principles of Genetics,* McGraw-Hill Book Company, Inc., New York, 1950.)

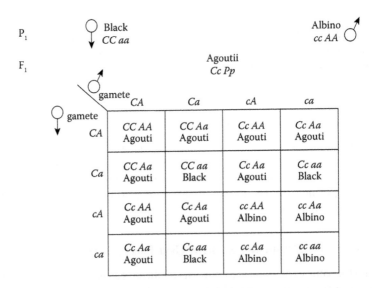

FIGURE 10.8 This checkerboard shows 9:3:4 ratio showing presence of *C* is necessary to express any other color. The gene *C* produces black-color mice when the albino gene is in recessive condition. (Redrawn from Sinnott, E.W., Dunn, L.C., and Dobzhansky, Th., *Principles of Genetics,* McGraw-Hill Book Company, Inc., New York, 1950.)

- In soybean, PI 200492 contains dominant gene (*Rpp1*) for resistance to soybean rust.
- 12ST4-5 line derived from *Glycine max* cv. Dwight (*2n*= 40) × *G. tomentella*, PI 441001 (*2n*= 78) is resistant to soybean rust.
- In F_2, seedlings of 12ST4-5 × PI 200492 segregated for 15 (resistant):1 (susceptible) ratio.
- This demonstrates that 12ST4-5 has a different gene to *Rpp1* found in PI 200492.

Thus, only recessive homozygous plants show a susceptible reaction.

10.5   TEST OF INDEPENDENCE

A clear understanding of the probability is of fundamental importance to understand the laws of heredity: (1) to appreciate the operation of genetic mechanism, (2) predicting the likelihood of certain results from particular hybridizations, and (3) determining how well an F_1 phenotypic ratios fit a particular postulated genetic mechanism. Based on transmission of two pairs of genes which are assumed to be on separate chromosome pairs, the chance of the simultaneous occurrence of two or more independent events is equal to the product of the probability that each will occur separately. The application of the probability to genetics permits demonstration of the two independent, nongenetic events, the binomial expression, and to determine the goodness of fit.

10.5.1   TWO INDEPENDENT NONGENETIC EVENTS—TWO-COIN TOSSES

This can be elucidated by tossing two coins simultaneously 50 times (Burns 1969). The outcome of any single toss will be HH (head), HT (head and tail), and TT (tail). Of the 50 tosses, the results were:

HH	12
HT	27
TT	11

We can assume three possibilities: (1) the deviation from the predicted result is within limits set by chance alone and the fall of coins is unbiased; (2) if two coins are tossed simultaneously, does each one of them have an equal chance of coming to rest heads or tails? In this case, there is a remote chance that the ultimate fall of one coin affects the other. Thus, the second hypothesis will be that the coins themselves are independent of each other; and (3) the third assumption may be, at least at the outset, is that successive tosses are also independent of each other. This would tell that having obtained two heads the first toss does not in any way affect the outcome of the second toss or any other tosses.

We expect to see 1/4 HH + 1/2 HT + 1/4 TT; thus, for two independent events of known probability it can be stated as $a^2 + 2ab + b^2$. The *a* representing head and *b* representing tail and can be simplified as $(a + b)^2$. We can summarize the observed and expected values of tossing two coins of head and tail:

Class	Observed	Expected
HH	12	12.5
HT	27	25.5
TT	11	12.5
	50	50.0

10.5.2 FOUR INDEPENDENT NONGENETIC EVENTS—FOUR-COIN TOSSES

In four-coin tosses, the possible combinations of heads and tails actually obtained in an experiment are as follows:

Class	Observed
HHHH	9
HHHT	32
HHTT	29
HTTT	25
TTTT	5
	100

In a two-coin toss, the expansion of $(a + b)^2$ gives an expected ratio of $a^2 + 2ab + b^2$. Therefore, by expending $(a + b)^4$, we expect: $a^4 + 4a^3b + 6\,a^2b^2 + 4ab^3 + b^4$. By substituting the numerical values, we will obtain:

$$(1/2)^4 + 4[(1/2)^3 \cdot 1/2] + 6[(1/2)^2 \cdot (1/2)^2] + 4[(1/2) \cdot (1/2)^3] + (1/2)^4 = 1 \text{ or}$$
$$1/16 + 4/16 + 6/16 + 4/16 + 1/16 = 1$$

We can compare observed results with the calculated results for expectancy in a four-coin toss:

Class	Observed	Calculated	
HHHH	9	6.25	(=1/16 of 100)
HHHT	32	25.00	(=4/16 of 100)
HHTT	29	37.50	(=6/16 of 100)
HTTT	25	25.00	(=4/16 of 100)
TTTT	5	6.25	(=1/16 of 100)
	100	100.00	

The chi-square test is used to accept the observed and expected assumptions. The formula of chi-square is:

$$\chi^2 = \Sigma\,[(o\text{-}e)^2]\,e$$

o = observed frequencies

c = calculated frequencies

Σ = summed for all classes

The calculated chi-square of coin tosses (1:2:1 expectation) can be expressed as:

Class	Observed O	Expected e	Deviation o-e	Squared Deviation $(o-e)^2$	$\dfrac{(o-e)^2}{e}$
HH	12	12.5	−0.5	0.25	0.02
HT	27	25.0	+2.0	4.00	0.16
TT	11	12.5	−1.5	2.25	0.18
	50	50.0	0		$\chi^2 = 0.36$

How often, by chance, can we expect a value of $\chi^2 = 0.36$ in a 1:2:1 ratio? Thus, we can accept that the observed deviation was produced only by chance. To answer this question, we should consult a table of chi-square (Table 10.2). In order to use this table, it is necessary to know the degrees of freedom; it is one less than the number of classes involved and represents the number of independent classes for which a chi-square value is calculated. Based on probability values and the degree of freedom presented in Table 10.2, we can conclude that the observed chi-square 0.36 with two degrees of freedom fits probability values between 0.95 and 0.80. This suggests that for an expected ratio of 1:2:1, we can expect a deviation as large as, or larger than we experienced, in between 80% and 95%. Such deviation could be due to chance and our assumption is good, and a good fit between observed results and our expected values.

TABLE 10.2
Chi-Square Table

Degrees of Freedom	Probability										
	0.95	0.90	0.80	0.70	0.50	0.30	0.20	0.10	0.05	0.01	0.001
1	0.004	0.02	0.06	0.15	0.46	1.07	1.64	2.71	3.84	6.64	10.83
2	0.10	0.21	0.45	0.71	1.39	2.41	3.22	4.60	5.99	9.21	13.82
3	0.35	0.58	1.01	1.42	2.37	3.66	4.64	6.25	7.82	11.34	16.27
4	0.71	1.06	1.65	2.20	3.36	4.88	5.99	7.78	9.49	13.28	18.47
5	1.14	1.61	2.34	3.00	4.35	6.06	7.29	9.24	11.07	15.09	20.52
6	1.63	2.20	3.07	3.83	5.35	7.23	8.56	10.64	12.59	16.81	22.46
7	2.17	2.83	3.82	4.67	6.35	8.38	9.80	12.02	14.07	18.48	24.32
8	2.73	3.49	4.59	5.53	7.34	9.52	11.03	13.36	15.51	20.09	26.12
9	3.32	4.17	5.38	6.39	8.34	10.66	12.24	14.68	16.92	21.67	27.88
10	3.94	4.86	6.18	7.27	9.34	11.78	13.44	15.99	18.31	23.21	29.59
	Nonsignificant								Significant		

Source: R.A. Fisher and F. Yates, *Statistical Tables for Biological, Agricultural and Medical Research.* 6th ed., Table IV, Oliver & Boyd, Ltd., Edinburgh, 1963, by permission of the authors and publishers.

The chi-square test is very useful for obtaining an objective approximation of goodness of fit. However, this is reliable when the observed and expected frequency in any class is five or more. The method provides useful information for an earlier judgmental answer.

10.6 GENETIC MAPPING OF CHROMOSOMES

10.6.1 CLASSICAL METHOD

Walter Stanborough Sutton (1903) discovered that chromosomes must carry genes that are inherited. In case two genes do not follow Mendelian inheritance, these genes are lined up and on the same chromosomes. Bateson and Punnett (1905–1908; see Peters 1959) could not explain, in sweet pea (*Lathyrus odoratus*), the following F_2 results:

P_1	Purple flowers and long pollen grains (*R Ro/R Ro*) × red flowers and round pollen grains (*r ro/ r ro*)
F_1	All purple long (*R Ro/ r ro*)
F_2	

Traits	Observed	Expected (7:1:1:7)
Purple long (*R Ro/ r ro*)	296	295
Purple round (*R Ro/r ro*)	19	25
Red long *r ro/Ro Ro*)	27	25
Red round (*r ro/r ro*)	85	82
	427	427

These results do not fit the 9:3:3:1 ratio as χ^2 turned out to be 219.27. Thus, the segregation ratio is not acceptable. Bateson and Punnett could not figure out the cause of this segregation and expected a ratio of 7:1:1:7 (above). However, now it is known that these two genes are linked. These genes may be arranged in a chromosome in two ways: (1) The two dominants, *R* (purple) and *Ro* (long) may be located on one member of the chromosome pair and two recessive, *r* and *ro*, on the other chromosome pair. This arrangement is known as *cis*. (2) The dominant of one pair and the recessive of the other may be located on one chromosome of the pair, with the recessive of the first gene pair and dominant of the second gene pair on the other chromosome. This arrangement is known as *trans* (Figure 10.9).

The classical genetic maps using qualitative genes are based on the extent of recombination frequencies observed either from the testcross or by selfing the F_1 plants. In testcross, F_1 plants are backcrossed to the recessive homozygous parent and progenies should segregate in a ratio of 1:1:1:1 if genes are more than 50 cM apart on the same chromosome or on different chromosomes. A departure from this ratio suggests that two genes are linked and frequency depends upon how close are

FIGURE 10.9   Genes arranged in *cis* and *trans* positions. (Redrawn from Burns, G.W., *The Science of Genetics. The Macmillan Biology Series*, The Macmillan Company, Collier Macmillan Limited, London, pp. 103, 1969.)

the genes. If Bateson and Punnett would have conducted testcross, the following segregation is expected:

P	Homozygous purple flowers long pollen grains (*R Ro/R Ro*) × red flowers round pollen grains (*r ro/ r ro*)	
F_1	All purple long (*R Ro/ r ro*)	
Testcross progeny	Purple long (*R Ro/ r ro*)	= 192
	Red round (*r ro/r ro*)	= 182
	Purple round (*R Ro/r ro*)	= 023
	Red long *r ro/Ro Ro*)	= 030
		427

These results depart widely from the 1:1:1:1 ratio. Two groups resemble parents while two groups, with far less frequencies, are nonparental types and this deviation is too large than the expected 1:1:1:1 ratio. To check this hypothesis, the chi-square test for a 1:1:1:1 expectation shows:

Phenotype	o	e	o-e	(o-e)²	(o-e)²/e
Purple long	192	106.75	85.25	7267:563	68.08
Red round	182	106.75	75.25	5662.563	53.05
Purple round	23	106.75	−83.75	7014.063	65.70
Red long	30	107.75	−76.75	5890.563	55.91
	427	427.00	0.00		$\chi^2 = 242.01$

Testcross progeny: The following table of testcross progenies shows that 87.6% of plants are parental types and 12.4% are nonparental types.

Phenotype	Genotype	Number	Frequency	Type
Purple long	*R Ro/r ro*	192	0.4496	Parental
Red round	*r ro/ r ro*	192	0.4262	Parental
Purple round	*R ro/r ro*	23	0.0538	Nonparental
Red long	*r Ro/r ro*	30	0.0702	Nonparental
		427	1.0000	

If R and Ro genes are not linked, it is expected to have 50% each of parental and nonparental types because parents should have produced gametes in equal frequency. Bateson and Punnett recognized that their population was segregating as seven purple long, one purple round, one red long, and seven red round. However, it should have segregated in a ratio of 9:3:3:1 if genes are unlinked.

10.6.2 CHROMOSOME MAPPING

10.6.2.1 Crossing Over and Crossover Frequency

If genes are arranged in *cis* linkage, it can be easily determined based on the frequency of the four possible types of F_1 gametes in unequal frequency. Depending upon the closeness of genes on the same chromosome, the occurrence of chiasma during meiotic prophase-I and separation of chromosomes at anaphase-I, suggest that 87.6% gametes are contained in the parental type chromosomes and 12.4% in crossover (nonparental chromosome type). The more closely the genes, the frequency of crossover product is reduced and is described as a map unit. A map unit is equal to 1% of crossing over; this represents the linear distance within which 1% crossing over takes place. Thus, for sweet pea, the distance from R and Ro would be 12.4 map units or cM.

Classical chromosome mapping is easily conducted based on the crossover frequencies observed from the testcross, rather than the data observed from the F_2 population. The association of genes using dihybrid testcrosses in *Drosophila melanogaster*, the fruit fly, has been well known; normal is designated as +:

Gene Symbol	Phenotypes
+	Normal wing (dominant)
cu	Curled wing (recessive)
+	Normal thorax (dominant)
sr	Striped thorax (recessive)
+	Normal bristles (dominant)
ss	Spineless bristles (recessive)

Testcross 1 in *cis* arrangement:

P_1	Female		Male
	Normal wing normal thorax	×	Curled wing striped thorax
	+ + / *cu sr*		*cu sr* / *cu sr*

Progenies:

F_1 Phenotypes	Maternal Chromosomes	Numbers	Percentage
Normal normal	+ +	436	43.6 parental
Curled wing and striped	*cu sr*	468	46.8 parental
Normal striped thorax	+ *sr*	42	4.2 crossover
Curled wing normal	*cu* +	54	5.4 crossover

Parental types: 90.4%; crossover types: 9.6%

This suggests that *cu* and *sr* are linked and the distance between two genes is 9.6 cM.

Testcross 2 in *trans* arrangement:

P$_1$	Female	Male
	Normal wing normal thorax ×	Curled wing striped thorax
	+ sr / cu /+	cu s r / cu sr

Progenies:

F$_1$ Phenotypes	Maternal chromosomes	Numbers	Percentage
Normal normal	+ +	49	4.9 parental
Curled wing and striped	cu sr	47	4.7 parental
Normal striped thorax	+ sr	442	44.2 nonparental
Curled wing and normal	cu +	462	46.2 nonparental

Parental types: 9.6%; nonparental types: 90.4%

It should be noted that the percentage of recombinant is 9.6%, but with the *trans* linkage in the female parent, the parental types in the F$_1$ now constitute the smaller class.

Testcross 3 in *trans* arrangement:

P$_1$	Female	Male
	Normal wing and normal bristles ×	Curled wing and spineless bristles
	+ + / cu ss	cu ss / cu ss

Progenies:

F$_1$ Phenotype	Maternal Chromosomes	Numbers	Percentage
Normal wing and normal bristles	+ +	450	45.0 parental
Curled wing and spineless bristles	cu su	465	46.5 parental
Normal wing and spineless bristles	+ ss	39	3.9 nonparental
Curled wing and normal bristles	cu +	46	4.6 nonparental

Parental types: 91.5%; nonparental types: 8.5%

The *trans* arrangement in the female gives compatible results:
Testcross 4 in *cis* arrangement:

P$_1$	Female	Male
	Normal normal ×	Curled wing and spineless bristles
	+ ss / cu +	cu ss / cu ss

F₁ Phenotypes	Maternal Chromosomes	Numbers	Percentage
Normal wing and normal bristles	+ +	45	4.5 parental
Curled wing and spineless bristles	cu ss	40	4.0 parental
Normal wing and spineless bristles	+ ss	461	46.1 nonparental
Curled wing and normal bristles	cu +	454	45.4 nonparental

Parental types: 8.5%; nonparental types: 91.5%

At this stage, we have the following distance cM among the genes based on crossing over and this can be represented as follows:

$$cu\text{-}sr \quad 9.6$$
$$cu\text{-}ss \quad 8.5$$

Based on the above information, we can assume that cu, ss, and sr genes are on the same chromosomes and their arrangement may be either a or b. To determine the correct alternatives, we should know the crossover frequencies between ss and sr which are 9.6 + 8.5 = 18.1 or 9.6 − 8.5 = 1.1.

10.6.2.2 Three-Point Testcross

The three-point testcross provides more precise distances among three genes than those obtained from a two-point testcross (Figure 10.10a). In *Drosophila*, three pairs of genes: a recessive homozygote cu, ss, and sr fly were testcrossed with a fly with dominant (normal) phenotype. Based on segregation in testcross progenies we can determine the correct gene sequence.

P₁	Female	Normal Normal Normal × Curled Spineless Striped	Male
		+ + +/cu ss sr	cu ss sr/ cu ss sr

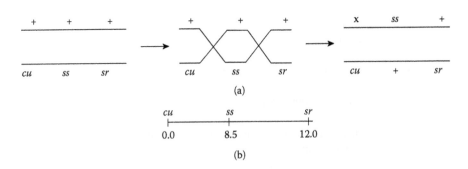

(a)

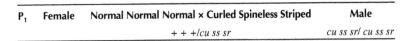

(b)

FIGURE 10.10 (a) Double crossing over; (b) Arrangement of genes on the chromosome. (Redrawn from Burns, G.W., *The Science of Genetics. The Macmillan Biology Series,* The Macmillan Company, Collier Macmillan Limited, London, pp. 113, 1969.)

F1 Phenotype	Maternal Chromosomes	Number	%	Type
Normal normal normal	+ + +	440	44.0	Parental
Curled spineless striped	cu ss sr	452	45.2	Parental
Normal spineless striped	ss sr+	40	4.0	cu-ss single c.o.
Curled normal normal	cu + +	33	3.3	cu-ss single c.o.
Normal normal striped	+ + sr	11	1.1	ss-sr single c.o.
Curled spineless normal	cu ss +	12	1.2	ss-sr single c.o.
Normal spineless normal	+ ss +	7	0.7	Double c.o.
Curled normal striped	cu + sr	5	0.5	Double c.o.

Parental: 89.2%
Single c.o. (cu-ss): 7.3%
Single c.o. (ss-sr): 2.3%
Double c.o.: 1.2%

Based on data observed in the above table we can establish the following conclusions:

a. *Double crossover*: The double crossover frequency is the smallest (0.7%; 0.5%) and may determine the gene sequence. Since genes in the female parent are in *cis* arrangement, then + *ss* + and *cu* + *sr* individuals must be the product of double crossover.

b. Thus, the gene sequence on the map will be *cu-ss-sr* (Figure 10.10a and b).

c. The precise distance between *cu* and *ss* is 7.0 + 1.2 = 8.5 (single crossover + double crossover) (Figure 10.10b).

d. The true distance between *ss* and *sr* is 2.3 + 1.2 = 3.5 (single crossover + double crossover).

e. The correct distance between *cu* and *sr* is 7.3 +1.2 + 2.3 + 1.2 = 12.0 (*cu-ss* single crossover; *ss-sr* single crossover + twice the double crossovers). The double crossover implies the crossing over between *cu* and *ss* and between *ss* and *sr*. In two-point crossover, the distance between *cu* and *sr* is 9.6, lower than those observed in a three-point test cross (12.0) (Figure 10.10b). This is due to the inability to detect double crossovers without a third marker between *cu* and *sr*.

f. It has been established based on classical cytogenetics that the number of known linkage groups never exceeds the number of pairs of homologous chromosomes in diploid organisms.

g. Assuming this is the first linkage group, three genes can be placed as shown in Figure 10.11a and b.

h. In *Drosophila*, when the female homozygote fly (*cu ss sr/cu ss sr*) is crossed using F_1 heterozygote (+ + + / *cu ss sr*) as a male, only two parental phenotypes will be recovered because crossing over does not occur in male flies; dipterans are unusual in this aspect.

10.6.2.3 Interference and Coincidence

Once a crossing over occurs in one chromosome segment, the probably of another crossing over in the adjacent region is reduced. This phenomenon is known as interference.

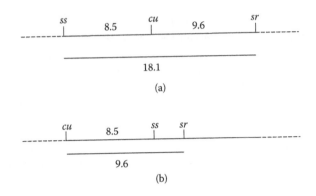

FIGURE 10.11   (a) Arrangement and order of genes on the chromosomes. (b) Change in the order after detecting double crossover. (Redrawn from Burns, G.W., *The Science of Genetics, The Macmillan Biology Series*, The Macmillan Company, Collier Macmillan Limited, London, pp. 111, 1969.)

Interference appears to be unequal in different parts of a chromosome and among chromosomes of a given complement. In general, interference appears to be greatest near the centromere and at the distal ends of a chromosome. The degrees of interference are commonly expressed as the coefficient of coincidence. Coincidence values ordinarily vary from 0 to 1; absence of interference results in a coincidence value of 1, whereas complete interference results in a coincidence of 0. Coincidence is usually quite small for a short map distance.

   In conclusion, knowledge of Mendelian genetics and its modified ratios is the foundation of the modern genetics, cytogenetics, plan breeding, and molecular biology.

10.7   RISE AND DECLINE OF PLANT CYTOGENETICS

Shortly after the rediscovery of Mendel's laws of inheritance, Walter Sutton (1903) proposed the chromosome in heredity based on studies in *Brachystola* (grasshopper):

- The chromosome group of the presynaptic germ cells is made up of two equivalent chromosome series, and strong grounds exist for the conclusion that one of these is paternal and the other maternal.
- The process of synapsis (pseudoreduction) consists of the union in pairs of the homologous numbers (i.e., those that correspond in size) of the two series.
- The first postzygotic or maturation mitosis is equational and hence results in no chromosome differentiation.
- The second postzygotic division is a reducing division, resulting in the separation of the chromosomes which have conjugated in synapsis, and their relegation to different germ cells.
- The chromosomes retain a morphological individuality throughout the various cell divisions.

- Based on the above observations, Sutton demonstrated that chromosomes are the vehicle for the genes used by Mendel in sweet peas.
- Thus, cytogenetics is a hybrid of cytology and genetics. Several organisms have been used in cytogenetic studies including *Drosophila* (Morgan, Sturtevant, Bridges 1910–1922), *Datura* (Blakeslee, Avery, Satina, and Rietsema and associates [1912–1956]), maize (McClintock, Randolph, Rhoades, and others), wheat (Kihara and colleagues, Sears and colleagues, and Riley and colleagues), barley (Tsuchiya and colleagues), tomato (Rick, and Khush and colleagues), and others.
- Cytogenetic chromosome maps for barley, tomato, rice, and maize have been developed by using aneuploidy stocks such as primary trisomics. Unfortunately, when these pioneering cytogeneticists either retired or expired, their positions were filled by molecular geneticists designating a new field, molecular cytogenetics.

Aneuploid germplasm stocks are being deposited and maintained in various ARS repositories (http://www.ars-grin.gov/npgs/rephomepgs.html). The Center for Agricultural Resources Research (CAAR) (http://www.ars.usda.gov/main/site_main.htm?modecode=30-12-30-00) located in Fort Collins, Colorado, preserves one set of plant and animal genetic resources.

However, active classical cytogenetic research is minimal and sometimes materials are lost; for example, primary trisomics of soybean is unavailable although results are in publications. Therefore, classical cytogenetics is dying a slow death. Cytogenetics is being taught by plant breeders without a cytology laboratory. My personal experience is that students are graduating in plant breeding and genetics without seeing the chromosomes.

REFERENCES

Blixt, S. 1975. Why didn't Gregor Mendel find linkage? *Nature* 256: 206.
Burns, G. W. 1969. *The Science of Genetics. The Macmillan Biology Series*. N. H. Giles, and J. G. Torrey (eds). The Macmillan Company, Collier Macmillan Limited, London, pp. 399.
Peters, J. A. 1959. *Classic Papers in Genetics*. Prentice-Hall, Englewood Cliffs, NJ.
Sinnott, E. W., L. C. Dunn, and Th. Dobzhansky. 1950. *Principles of Genetics*. McGraw-Hill Book Company, Inc. New York.
Strickberger, M. W. 1968. Genetics. The Macmillan Company, New York.
Sutton, W. S. 1903. The chromosomes in heredity. *Biological Bulletin* 4: 231–251.

11 Structural Chromosome Changes for Locating the Genes

11.1 DEFICIENCIES

- The loss of a segment from a normal chromosome is known as deficiency.
- Deficiency refers to any chromosome loss and may be terminal (Figure 11.1a and b) or intercalary (Figure 11.1c and d) (Khush and Rick 1968).
- The recessive gene expresses in both types; it is known as pseudodominant and has been effectively utilized for locating qualitative genes in the chromosomes of maize and tomato.

11.1.1 ORIGIN OF DEFICIENCIES

- Deficiencies in chromosomes of diploid plants are rarely identified, however, they have been induced by irradiating pollen grains of normal plants by X-rays and fast neutrons.
- The irradiated pollens were pollinated onto recessive homozygous morphological mutants.

11.1.2 IDENTIFICATION OF DEFICIENCIES

- Chromosome pairing at pachynema stage.
- Appearance of recessive gene in crosses from mutant female and X-rayed pollen, known as pseudodominant technique.

11.1.3 GENETIC STUDIES

- McClintock (1938) pollinated silk of *bm 1* plant with X-rayed pollen of maize containing a normal haploid complement with dominant gene *Bm 1*. Of the 466 plants, two plants were variegated for *Bm 1* and *bm 1*. This aberrant behavior of the ring chromosome was produced by the X-ray treatment that carried the *Bm 1* gene. The loss of ring chromosome expresses the recessive variegated trait.
- Khush and Rick (1968) examined 74 deficiencies of tomato chromosomes induced by X-rays and fast neutrons and identified deficiencies by a pseudododominant technique using pachytene chromosomes. They identified 35 genes of the 18 (chromosome 1, 2, 3, 4, 6, 7, 8, 9, 10, 11, and 12) of the 24 arms of the complement.

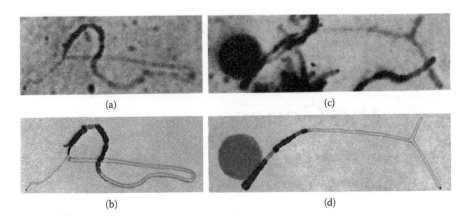

FIGURE 11.1 Pachynema chromosome of tomato showing terminal deficiency of 1S; (a) photomicrograph and (b) showing interpretive drawing. Breakage occurred in the heterochromatic region. (c) Photograph showing interstitial deficiency of 2L. The deficiency loop is folded to pair nonhomologously; (d) shows an interpretive drawing; breakage occurred in the euchromatic region. (From Khush, G.S., and Rick, C. M., *Chromosoma*, 23, 452–484, 1960.)

11.2   DUPLICATIONS

An extra chromosome segment is attached to either the same homologous chromosome or transposed to one of the nonhomologous members of the genome.

11.2.1   CLASSIFICATION OF DUPLICATIONS

Duplications are classified as:

1. Normal chromosome (Figure 11.2A)
2. Tandem duplication (Figure 11.2B)
3. Reverse tandem duplication (Figure 11.2C)
4. Duplication in a different arm (Figure 11.2D)
5. Nonhomologous (displace) duplication (Figure 11.2E)

11.2.2   ORIGIN OF DUPLICATIONS

- Duplications occur in nature; several diploid crops that duplicate genes, determined by the F_2 segregation (15:1 or 9:7) have been recorded.
- Usually, X-rays induce breakage in the chromosomes producing duplications and deficiencies.
- In maize, B–A translocations produce heritable, stable proximal and distal duplications (Carlson and Roseman 1991).

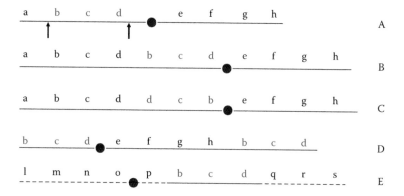

FIGURE 11.2 Diagrammatic representation of five types of possible duplications. (A)≈Normal chromosome, (B) tandem duplication, (C) reverse tandem duplication; (D)≈duplication in a different arm, (E) nonhomologous (displaced duplication).

11.2.3 GENETIC STUDIES

- Duplications can modify genetic ratios because of inviable gametes modifying phenotypes due to dosages effects of a particular allele.
- Distortion in genetic segregation ratios of linked genes may be used to determine the position of translocation points with respect to each other.

11.3 CHROMOSOMAL INTERCHANGES

- Interchanges are known as segmental chromosomal interchanges, reciprocal translocations, or simply translocations.
- Translocations are the result of the reciprocal exchange of terminal segments of nonhomologous chromosomes (Figure 11.3).

11.3.1 ORIGIN OF INTERCHANGES

- It occurs spontaneously in nature; the classical example of natural reciprocal translocation is in *Oenothera* and *Secale*.
- It is induced by X-ray treatment.

11.3.2 IDENTIFICATION

11.3.2.1 Cytological Method

- At the pachynema stage, a cross-shaped structure is observed (Figure 11.4).
- At metaphase-I, the majority of the sporocytes show 1IV + II.
- Alternate-1 association (zigzag) and migration at anaphase-I, homologous kinetochores $(1 + 2; 1^2 + 2^1)$ moves to opposite poles resulting in viable gametes, if there is no crossing over in the interstitial region.

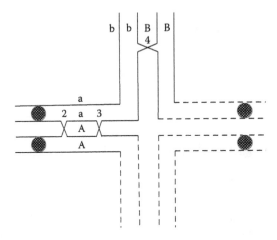

FIGURE 11.3 Diagrammatic sketch of a reciprocal translocation heterozygote at pachynema showing crossing overs between kinetochore and marker gene *a* (2), between marker gene *a* and breakage point (interstitial region-3) and between breakage point and marker gene *b* (distal region 4).

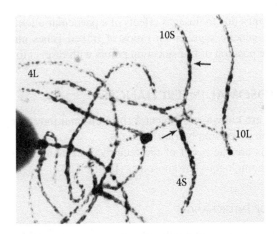

FIGURE 11.4 A photomicrograph of a cell at pachynema showing a cross-shaped configuration in an heterozygous reciprocal translocation in maize. Chromosomes involved in the interchange are numbered and kinetochores are shown by an arrow (Courtesy of Dr. D. F. Weber).

- In adjacent-1 $(1 + 2^1, 2 + 1^2)$, all gametes from the original combination, and crossing over in the distal region; between breakage point and marker gene b is viable but all gametes are nonviable. Only cytological and genetically crossover products are viable. By contrast, the adjacent-2 $(1^2 + 1, 2 + 2^1)$ arrangement produces all inviable gametes.
- An approximate 1:1 ratio of alternate and adjacent chromosome configurations at metaphase-I results in 50 % pollen fertility in plants with interchange heterozygotes.

- In a noncooriented arrangement, normal chromosomes carrying nonhomologous kinetochores are on the opposite side of a quadrivalent. Two noncooriented interchanged chromosomes are stretched in the middle and are not attached to the poles. This orientation results in 3 to 1 chromosome segregation.

11.3.2.2 Pollen Fertility

- Pollen and seed fertility: An interchange heterozygote in diploid crops is identified by partial pollen sterility in an F_1 plant. Pollen fertility is about 50%.
- In polyploid crops, interchanged heterozygote plants do not express 50% pollen sterility because Dp–Df spores are as competitive as normal pollen and female spore, and express normal fertility.

11.3.3 GENETIC STUDIES

- Interchange stocks have been used to identify linkage groups, to associate new mutants to specific chromosomes and to construct new karyotypes.
- The interchange gene-linkage analysis is based on the association of contrasting traits with partial sterility.
- An interchange behaves as a dominant marker for partial sterility located simultaneously in the two interchanged chromosomes at the points where the original breakage and exchange have occurred.

Translocated chromosome		T			
Normal chromosome		N			
Dominant marker gene		A			
Recessive gene		a			
F_1		$TNAa$			

Gametes	Male →	TA	NA	Ta	Na
	Female				
	↓				
F_2	TA	TTAA	TNAA	TTAa	TNAa
	NA	TNAA	NNAA	TNAa	NNAa
	Ta	TTAa	TNAa	TTaa	TNaa
	Na	TNAa	NNAa	TNaa	NNaa

- Normal, partial sterile, translocation heterozygote (TNAA and TNAa): 6
- Normal, fertile, translocation homozygote/normal homozygote: 6
- Mutant, partial sterile, translocation heterozygote: 2
- Mutant, fertile, translocation homozygote/normal homozygote: 2

- Examine plants with the recessive homozygotes in the F_2 population.
- TTaa (1) plants originated from Ta originated from the recombinant gametes.
- TNaa (2) originated from the union of recombinant and nonrecombinant gametes.

- NNaa (1) plants originated from nonrecombinant gametes.
- Thus, F_2 recessive homozygote should segregate for 1 (TT):2 (TN) and 1 (NN) plants but deviation from this can determine the linkage.
- In barley, glossy seedlings and virido-xantha are located on chromosome 4 (4H) because both mutants show linkage with T4-5e and independent segregation with T1-5a.
- Both mutants are close to a breakage point because no recombinant (N/T) type is observed.

11.3.4 Principles of Producing Interchange Testers

- The occurrence of two quadrivalents (IV) in F_1 plants suggests that both translocation testers involve different chromosomes.
- The presence of one hexavalent (VI) in F_1 sporocytes indicates that one of the chromosomes involved in the two translocation stocks is common.
- When both translocation stocks involve the same chromosomes, F_1 hybrids may show bivalents only.

11.4 TRISOMICS

- An organism containing a normal chromosome complement and one extra chromosome is known as a trisomic.
- The extra chromosome may be primary, secondary, tertiary, telocentric, acrocentric, or compensating (Figure 11.5).

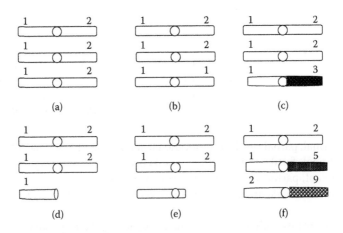

FIGURE 11.5 Diagrammatic sketch of various types of trisomics. (a) Primary trisomic; (b) secondary trisomic; (c) tertiary trisomic; (d) telotrisomic; (e) acrotrisomic; (f) compensating trisomic.

11.4.1 PRIMARY TRISOMICS

- An organism containing a normal chromosome complement and one extra chromosome is known as a primary trisomic (2x + 1) (Figure 11.6).
- Generally, autotriploids are considered to be one of the best and most dependable sources for establishing a primary trisomic series.
- The selfed progenies of autotriploid and autotriploid × diploid produce plants with $2n = 2x$, 2x +1, 2x + 2, 2x + 3, 2x + 4, 2x + 5, 2x + 6, 2x + 7, and 2x + 8 chromosomes.
- Asynaptic and desynaptic plants produce a low frequency of primary trisomics. These primary trisomics are not desired in genetic studies because they include sterile plants that cause the release of unrelated trisomics in their progenies.
- Primary trisomics have been isolated occasionally from the progeny plants treated with mutagens.
- Progenies of double trisomics, telotrisomics, secondary trisomics, tertiary trisomics throw related primary trisomics.

11.4.1.1 Identification of Primary Trisomics

11.4.1.1.1 Morphological Identification

- Initially, primary trisomics are identified and characterized based on modification of plant morphology (growth habit, plant height, degree of branching and tillering, leaf size and shape, color and texture of leaf surface, internode length, days to flower, seed fertility, and many other visible morphological traits) caused by the addition of an extra chromosome to the plant with diploid complement.
- They are compared with the diploid sibs.
- A complete primary trisomic series has been established in *Datura*, maize, barley, diploid oat, pearl millet, rice, *Arabidopsis*, and tomato.

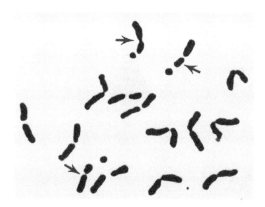

FIGURE 11.6 A photomicrograph of mitotic metaphase chromosomes of a primary trisomic ($2n = 15$) for Triplo 6 (purple) of barley showing three nucleolus organizer chromosomes (From Tsuchiya, T., *Jpn. J. Bot.* 17, 177–213, 1960.)

11.4.1.1.2   Cytological Identification

- The true identity of a primary trisomic is determined by cytological examination—mitosis and meiosis.
- They are designated based on length of chromosomes [Triplo 1, contains the longest chromosome, and so on].
- In barley, chromosomes 6 and 7 contain satellite chromosomes. In the primary trisomic for chromosome 6 (6H), mitotic metaphase cells show three satellite chromosomes (Figure 11.6).
- Pachytene chromosome analysis is a powerful tool to identify primary trisomics in maize, rice, and tomato.
- The three homologous chromosomes at pachynema compete to pair with one another, but only two of three homologues are synapsed end to end (normal fashion) while the third attempts to associate with the paired homologues in a random manner and form a loose trivalent configuration or pair itself (Figure 11.7).

11.4.1.1.3   Identification of Primary Trisomics with Translocation Testers

- The independence of primary trisomics can be verified by translocation tester sets.
- Use primary trisomics as a female parent and pollinate by the translocation testers.
- Analyze chromosome pairing in F_1 primary trisomics at diakinesis/metaphase-I.
- In critical combination, a chain of five chromosome (pentavalent) plus bivalent (Figures 11.8 and 11. 9) or the derivatives are obtained.

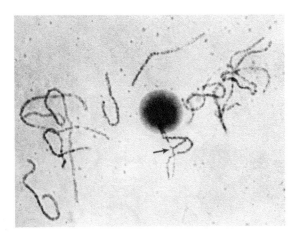

FIGURE 11.7 A trivalent configuration of Triplo 9 in rice at pachynema associated with the nucleolus. Arrow shows extra chromosome 9 partially paired nonhomologously. (From G.S. Khush et al., *Genetics* 107, 141–163, 1984.)

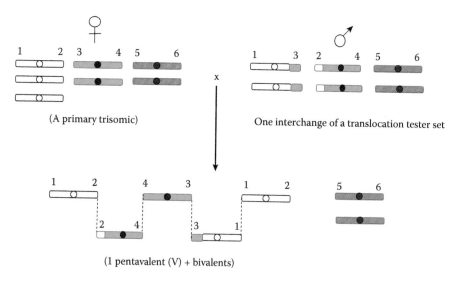

(A primary trisomic)

One interchange of a translocation tester set

(1 pentavalent (V) + bivalents)

FIGURE 11.8 Diagrammatic sketch showing procedure to identify the extra chromosome in primary trisomics by translocation tester sets.

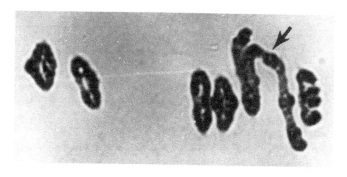

FIGURE 11.9 A meiotic metaphase-I chromosome configuration of 1V + 4II in an F_1 hybrid between primary trisomic bush and interchange suggests that the extra chromosome in bush is for chromosome 1. (From Singh, R.J., *Plant Cytogenetics*, 3rd ed., CRC Press, Boca Raton, FL, 2007.)

- In a noncritical combination, if the extra chromosome is not partly homologous with either of the interchanged chromosomes, a quadrivalent plus a trivalent plus bivalents are observed.

11.4.1.1.4 Genetic Segregation in Primary Trisomics

- Primary trisomics are very helpful for **locating a recessive gene** on a particular chromosome, verifying the independence of linkage group.
- The genetic ratios are modified from 3:1 (F_2) or 1:1 (BC_1) if a recessive gene is located on the extra chromosome (Figure 11.10).

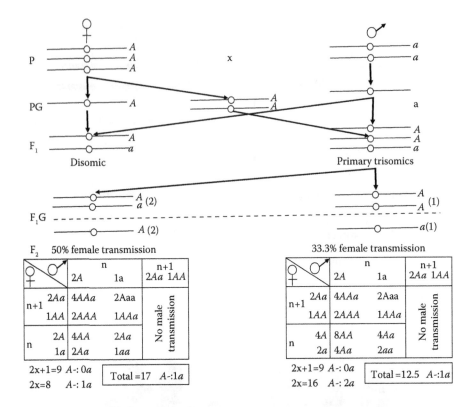

FIGURE 11.10 Diagrammatic sketch of random chromosome segregation showing association of a recessive marker (*a*) gene with a particular chromosome by primary trisomic method.

- In random chromosome segregation with 50% female transmission of (n + 1) primary trisomics, F_2 segregates for 17 *A*-: 1*a* ratio is expected.
- However, the expected 50% female transmission of n+1 in F_2 is not observed.
- Assuming 33.3% female transmission of n + 1 gametes in F_2, the phenotypic 17:1 ratio is modified to 12.5 *A*-:1*a*. (Figure 11.10).
- In random chromatid segregation, The F_2 ratio in 50 % female transmission of n + 1 gametes is 14*A*-:1*a* and in 33.3% female transmission of n+1 gametes it is 11.27*A*-:1*a*. A small frequency of recessive homozygous plants are identified.
- In case of locating a dominant gene on a particular chromosome, in random chromosome segregation, F_2 population segregates for 2*A*-:1*a* ratio.
- In case of locating a codominant gene on a particular chromosome, F_2 segregates for 6 (homozygous dominant):11 (heterozygous):1 (homozygous recessive) ratio.

11.4.2 SECONDARY TRISOMICS

- Secondary trisomic plants contain an extra isochromosome (both arms homologous) in addition to their normal somatic chromosome complement (Figure 11.5).

- Secondary chromosome (isochromosome) may arise directly or progressively by way of an unstable telocentric chromosome that would undergo a further misdivision or segregation without division.

11.4.2.1 Identification of Secondary Trisomics

11.4.2.1.1 Morphological Identification of Secondary Trisomics

- Morphologically, secondary trisomics for long arms modify plant morphology more drastically than those expressed by secondary trisomics for the short arms.
- In tomato, secondaries for long arms (6L, 7L, 9L, and 10L) showed much slower growth rates at all stages of plant growth than those recorded for the secondaries composed of short arms (3S, 7S, and 9S).
- Secondary trisomics for the short arms were indistinguishable morphologically from disomic (Khush and Rick 1969).

11.4.2.1.2 Cytological Identification

- Secondary trisomics show ring trivalent at diakinesis/metaphase-I.
- Secondary trisomics for the long arm show more ring trivalent than those for the short arms.

11.4.2.2 Genetic Segregation in Secondary Trisomics

- Secondary trisomics can be used to locate genes in a particular half-chromosome in much the same way as with primary trisomics.
- If a marker gene is located in the extra chromosome, a ratio of $3A$-:$1a$ :: all ($4A$-:$0a$) will be recorded for disomic and secondary trisomic fractions in random chromosome segregation.
- This suggests that no recessive homozygote plants will be obtained in a secondary trisomic fraction (Figure 11.11).
- In random chromatid segregation in F_2, the segregation ratios are:
  Disomic: $4.76A$-: $1a$
  Primary trisomic: $27.8A$-:$1a$
  Secondary trisomic: All A-:$0a$
- If a gene is not located on the secondary chromosome, disomic ratio ($3:1::\approx3:1$) in F_2 and BC_1 ($1:1::1:1$) are observed for diploid, primary trisomic, and secondary trisomic portions.

11.4.3 TERTIARY TRISOMICS

- A tertiary trisomic individual consists of an interchanged nonhomologous chromosome in addition to the normal somatic chromosome complement (Figure 11.5).
- Tertiary trisomics have been isolated from the progenies of translocation heterozygote.

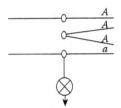

FIGURE 11.11 Diagrammatic sketch of random chromosome segregation showing association of a recessive gene (*a*) with a particular arm of a chromosome by use of secondary trisomics.

- An interchange heterozygote occasionally forms a noncooriented quadrivalent configuration at metaphase-I, and 3:1 random disjunction of chromosomes at anaphase-I will generate n+1 gamete.
- Thus, eight possible types of 2x + 1 individuals are expected in the progenies of a selfed interchange heterozygote. Of the four tertiaries, two will be in homozygous background and the other two in translocation heterozygous background.

11.4.3.1 Identification of Tertiary Trisomics

11.4.3.1.1 Morphological Identification

- Since tertiary chromosomes contain parts of two nonhomologous chromosomes, tertiaries inherit certain morphological features of the two related primaries.
- Tertiary trisomics carrying a long arm generally exert a greater influence on plant morphology than when carrying a short arm.
- Tertiaries for the short arms appear to contribute little or alter the phenotypic expression of the tertiary trisomics.

11.4.3.1.2 Cytological Identification

- In tertiary trisomics, a maximum association of five chromosomes (pentavalent) or derived configurations is observed.
- The formation of a pentavalent configuration depends upon the length of the tertiary chromosome.

11.4.3.2 Genetic Segregation in Tertiary Trisomics

- The genetic ratio in tertiary trisomics is modified the way it is in secondary trisomics, telotrisomics, and acrotrisomics.
- Assuming duplex genetic constitution (AAa) of an F_1, no male transmission of the tertiary chromosome, 50% female transmission, and a marker gene located on either arm of the tertiary chromosome, a trisomic ratio 3:1 ($2x$): all 4:0 ($2x + 1$) or a total of 7:1 ratio is expected.
- Furthermore, a 5:1 ratio should be obtained when the female transmission rate of the tertiary chromosome reaches 33.3%.
- If a marker gene is not located on the tertiary chromosome, a disomic ratio should be recorded in the F_2 for $2x$ and $2x + 1$ fractions.

11.4.4 TELOTRISOMICS

- A telocentric chromosome consists of a kinetochore (centromere) and one complete arm of a normal chromosome (Figure 11.5).
- A plant with a normal chromosome complement plus an extra telocentric chromosome is known as monotelotrisomic ($2x + 1$telocentric or telo).
- Monotelocentric chromosome originates spontaneously or in the progenies of primary trisomics.
- Misdivision of a centromere is a prerequisite and potent source for the production of telocentric chromosomes (Figure 11.12).

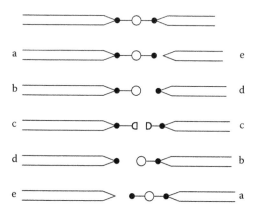

FIGURE 11.12 Diagrammatic sketch of origin of various types of telocentric chromosomes (Redrawn from Steinitz-Sears, L.M., *Genetics*, 54, 241–248, 1966. With permission.)

11.4.4.1 Identification of Telotrisomics

11.4.4.1.1 Morphological Identification

- In general, telotrisomic for the long arm of a chromosome has a similar effect on plant morphology as the corresponding primary trisomic.
- Usually, telotrisomics for the short arm are indistinguishable morphologically from diploid sibs.
- Occasionally, telotrisomics for certain chromosomes do not resemble primary trisomics. In rice, telotrisomics for chromosomes 1, 2, and 3 do not resemble diploid sibs or their respective primary trisomics.

11.4.4.1.2 Cytological Identification

- The telocentric chromosome is identified at mitotic metaphase or at meiotic pachynema.
- Somatic metaphase chromosome by acetocarmine staining identified Triplo 5L in barley (Figure 11.13).
- Giemsa N-banding verifies the acetocarmine staining and clearly showed that telocentric chromosome contains half of the kinetochore (Figure 11.14).
- In telotrisomic plants, a telocentric chromosome may associate with its normal homologues, forming a trivalent (Figure 11.15a and b) or remain a univalent.

11.4.4.2 Genetic Segregation in Telotrisomics

- Telotrisomics are useful in locating genes on a particular arm of a chromosome by a modified ratio in the F_2 population.

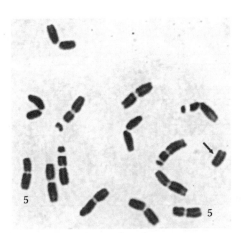

FIGURE 11.13 Mitotic metaphase chromosome of Triplo 5L after acetocarmine staining of barley. Compare long arm of chromosome 5 (chromosome 5 numbered) and Telo 5L (arrow). (From Singh, R.J. *Plant Cytogenetics,* 3rd edn., CRC Press, Boca Raton, FL, 2017.)

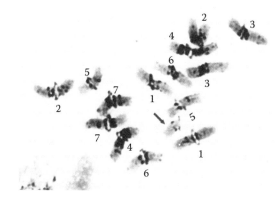

FIGURE 11.14 Giemsa N-banded mitotic metaphase chromosomes of Triplo 5L of barley. Triplo 5L contains half of the kinetochore (arrow) (From Singh, R.J., *Plant Cytogenetics,* 3rd ed., CRC Press, Boca Raton, FL, 2017.)

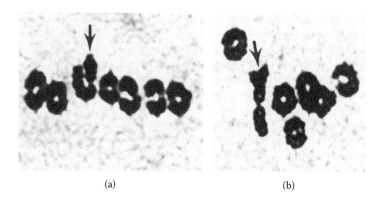

(a) (b)

FIGURE 11.15 Chromosome configurations at metaphase-I in Triplo 3L of barley. (a) Tandom V- shape trivalent. (b) Rod-shape trivalent; interstitial chiasma (arrow) between telocentric and normal chromosome is evident. Arrows show a telocentric chromosome. (From Singh, R.J., and T. Tsuchiya. *Bot. Gaz.* 142, 267–273, 1981. With permission.)

- If a gene is not located on the extra chromosome, a disomic ratio is obtained for diploid and trisomic fractions (3:1::3:1).
- If a gene is located on the extra telocentric chromosome, the F_2 population segregates as 3:1::4:0. No recessive homozygous plant will be obtained in the telotrisomic portion. If the chromosome count is not determined in the F_2, a ratio of 7:1 is expected.
- A 7A-:1a ratio (random chromosome segregation) expected of female transmission of telocentric is 50%. However, in 33.3% of female transmission of telocentric chromosomes, the ratio is modified to 5A-:1a (Figure 11.16).
- In random chromatid segregation, a 56.6B-:1b is observed in the telotrisomic portion and 4.76A-:1a in the disomic portion (Figure 11.17).

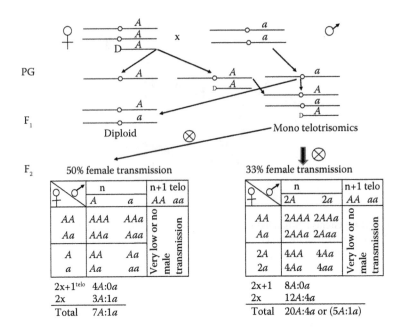

FIGURE 11.16 Diagrammatic sketch of random chromosome segregation showing association of a recessive gene (*a*) with a particular arm of a chromosome by use of telotrisomics.

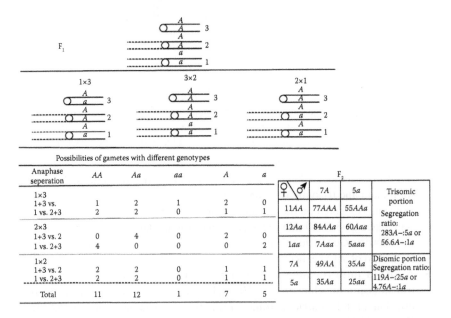

FIGURE 11.17 Diagrammatic sketch showing maximum equational chromatid segregation showing association of a recessive gene (*a*) with a particular arm of a chromosome by use of telotrisomics.

11.4.5 ACROTRISOMICS

- A plant containing an extra acrocentric chromosome is known as acrotri-somic (Figure11.5).
- Acrocentric chromosomes in barley may have originated by a break-age in chromosome arm(s), preferably adjacent to heterochromatic region (Figure 11.18).

11.4.5.1 Identification of Acrotrisomics

11.4.5.1.1 Morphological Identification

- Morphologically, acrotrisomics for the long arm are similar to their cor-responding primary trisomics.
- Acrotrisomics for the short arms are similar to disomic sib.

11.4.5.1.1 Cytological Identification

- Precise identification of acrocentric chromosome is conducted at mitotic metaphase by using acetocarmine and Giemsa N-banding techniques (Figure 11.19a and b).
- The extra acrocentric chromosome at diakinesis/metaphase-I remains as a univalent (Figure 11.20a) or associates with its normal homologues forming a trivalent configuration (Figure 11.20b).

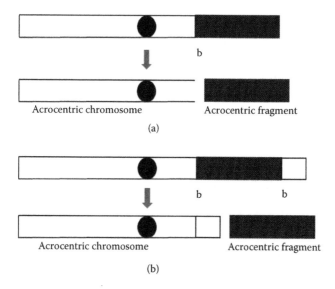

FIGURE 11.18 Diagrammatic sketch of the mode of origin of an acrocentric chromosome. (a) Single break at point b produces acrocentric chromosome with healed broken end and acentric fragment. (b) Two breaks at point b with the result of an acentric chromosome and the original telomere attached to form an acrocentric chromosome.

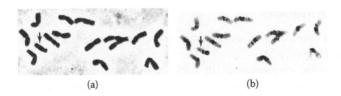

FIGURE 11.19 Acetocarmine (a) and Giemsa N-banded (b) acrotrisomic of a plant with $2n = 14 + 1$ acro3L^{3S}. Arrow shows an acro 3L^{3S}.

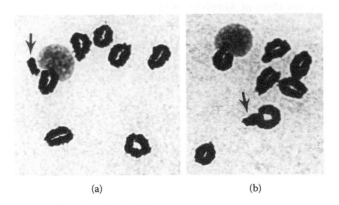

FIGURE 11.20 Chromosome configuration at diakinesis in plants with $2n = 14 + 1$ acro3L^{3S}. (a) 7II + 1I (arrow shows one univalent) (b) 6II +1III (ring and rod-shaped trivalent, arrow).

11.4.5.2 Genetic Segregation in Acrotrisomics

- The theoretical segregation ratios in acrotrisomic analysis are the same as those observed for telotrisomic analysis.
- Acrotrisomic linkage mapping is an excellent tool to locate genes precisely in a particular region of a chromosome.
- The theoretical segregation ratios in acrotrisomic analysis are the same as those observed for telotrisomic analysis.

REFERENCES

Carlson, W. R., and R. R. Roseman. 1991. Segmental duplication of distal chromosomal regions in maize. *Genome* 34: 537–542.

Khush, G. S., and C. M. Rick. 1968. Cytogenetic analysis of the tomato genome by means of induced deficiencies. *Chromosoma* 23: 452–484.

Khush, G. S., and C. M. Rick. 1969. Tomato secondary trisomics: Origin, identification, morphology, and use in cytogenetic analysis of the genome. *Heredity* 24: 129–146.

Khush, G. S., R. J. Singh, S. C. Sur, and A. L. Librojo. 1984. Primary trisomics of rice: Origin, identification, cytology and use in linkage mapping. *Genetics* 107: 141–163.

McClintock, B. 1938. The production of homozygous deficient tissues with mutant characteristics by means of the aberrant mitotic behavior of ring-shaped chromosomes. *Genetics* 23: 315–376.

Singh, R. J. 2007. *Plant Cytogenetics*. 3rd edn. CRC Press, Boca Raton, FL.

Singh, R. J., and T. Tsuchiya. 1981. Cytological study of the telocentric chromosome in seven monotelotrisomics of barley. *Bot. Gaz.* 142: 267–273.

Steinitz-Sears, L. M. 1966. Somatic instability of telocentric chromosomes in wheat and the nature of the centromere. *Genetics* 54: 241–248.

Tsuchiya, T. 1960. Cytogenetic studies of trisomics in barley. *Jpn. J. Bot.* 17: 177–213.

12 Wide Hybridization

12.1 INTRODUCTION

- Exotic germplasm (the alien species), relative to cultigens, is a rich reservoir for economically valuable traits and has been proven to be a useful source for increasing genetic variability in many crops by conventional plant breeding and cytogenetic methods.
- Exotic germplasm includes all germplasm that does not have immediate usefulness without hybridization and selection for adaptation for a given region.
- Alien genes are introgressed into cultigens by hybridization followed by chromosome doubling and by several backcrossings to the recurrent parent.
- Crop improvement requires a multidisciplinary approach (Figure 12.1).

12.2 GENE POOL CONCEPT

- A prerequisite for the exploitation of wild species to improve cultivars is to have complete comprehension and understanding of the taxonomic and evolutionary relationships between a cultigen and its wild allied species.
- Harlan and deWet (1971) proposed the concept of three gene pools based on hybridization among species.
  - Primary gene pool (GP-1): The primary gene pool consists of biological species. Hybridization within this gene pool is easy. F_1 plants are vigorous and fertile. In soybean (*Glycine max*), soybean cultivars, land races, and its wild annual progenitor *Glycine soja* are included in GP-1 (Figure 12.2).
  - Secondary gene pool (GP-2): The secondary gene pool includes all species that can be crossed with GP-1 species with at least some fertility in F_1. Soybean does not have GP-2. Wheat has a large number of species including *Aegilops*, *Secale*, and *Haynaldia*. Rice, cotton, and common bean has substantial GP-2.
  - Tertiary gene pool (GP-3): The tertiary gene pool (GP-3) is the extreme outer limit of potential genetic resource. Hybrids between GP-1 and GP-3 are difficult to produce, requiring *in vitro* technique to rescue F_1 plants; F_1 plants are anomalous, lethal, or completely sterile. All the wild perennial *Glycine* species belong to GP-3 for the soybean.

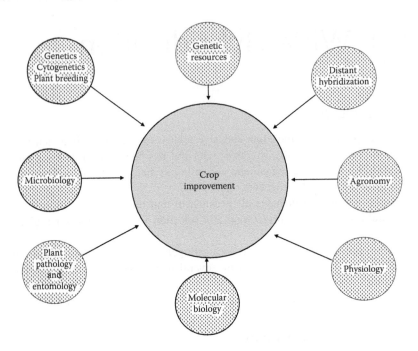

FIGURE 12.1 Diagrammatic sketch of a multidisciplinary approach for crop improvement. (From Singh, R.J., *Plant Cytogenetics*, 3rd ed., CRC Press, Boca Raton, FL, 2017.)

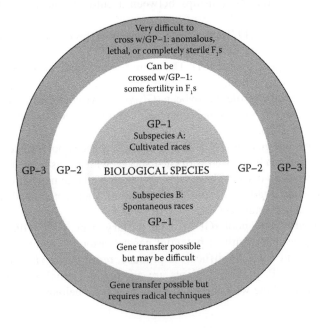

FIGURE 12.2 Gene pool concept in plants established based on hybridization. (Redrawn from Harlan, J.R. and de Wet, J.M.J., *Taxon,* 20, 509–517, 1971.)

12.3   UTILIZATION OF THE TERTIARY GENE POOL

Cytogeneticists use the following methods to overcome the hybrid barrier to utilize the genetic resources of GP-3.

- The assemblage of large germplasm
- Application of growth hormones to reduce the embryo abortion
- Immature seed culture rescue
- Improved culture conditions
- Restoration of the seed fertility by doubling the chromosomes of sterile F_1 plants
- Utilization of bridge crosses where direct crosses are not successful

12.4   THE ASSEMBLAGE OF LARGE GERMPLASM

- The conservation of crop diversity has become mandatory because adaptation of high-yielding cultivars and genetically modified crops are taking over the niches of wild annuals and perennial species.
- Fifteen international centers have been established, shown below, and these are under CGIAR (*the Consultative Group for International Agricultural Research*):
  - Africa Research Center (http://www.cgiar.org/cgiar-consortium/research-centers/africarice/)
  - Biodiversity International (http://www.cgiar.org/cgiar-consortium/research-centers/bioversity-international/)
  - Center for International Forestry Research (CIFOR) (http://www.cgiar.org/cgiar-consortium/research-centers/center-for-international-forestry-research-cifor/)
  - International Center for Agricultural Research in the Dry Areas (ICARDA) http://www.cgiar.org/cgiar-consortium/research-centers/international-center-for-agricultural-research-in-the-dry-areas-icarda/)
  - International Center for Tropical Agriculture (CIAT) (http://www.cgiar.org/cgiar-consortium/research-centers/international-center-for-tropical-agriculture-ciat/)
  - International Crop Research Institute for the Semi-Arid Tropics (ICRISAT) (http://www.cgiar.org/cgiar-consortium/research-centers/international-crops-research-institute-for-the-semi-arid-tropics-icrisat/)
  - International Food Policy Research Institute (IFPRI) (http://www.cgiar.org/cgiar-consortium/research-centers/international-food-policy-research-institute-ifpri/)
  - International Institute of Tropical Agriculture (IITA) http://www.cgiar.org/cgiar-consortium/research-centers/international-institute-of-tropical-agriculture-iita/)
  - International Livestock Research Institute (ILRI) http://www.cgiar.org/cgiar-consortium/research-centers/international-livestock-research-institute-ilri/)

- International Maize and Wheat Improvement Center (CIMMYT) http://www.cgiar.org/cgiar-consortium/research-centers/international-maize-and-wheat-improvement-center-cimmyt/)
- International Rice Research Institute (IRRI) (http://www.cgiar.org/cgiar-consortium/research-centers/international-rice-research-institute-irri/)
- International Potato Center (CIP) (http://www.cgiar.org/cgiar-consortium/research-centers/international-potato-center-cip/)
- International Water Management Institute (IWMI) (http://www.cgiar.org/cgiar-consortium/research-centers/international-water-management-institute-iwmi/)
- World Agroforestry Centre (ICRAF) (http://www.cgiar.org/cgiar-consortium/research-centers/world-agroforestry-centre/)
- World Fish (http://www.cgiar.org/cgiar-consortium/research-centers/worldfish/)

All international centers are unique and have different mandates. However, CIMMYT and IRRI have modified wheat and rice, respectively, that produced high-yielding wheat and rice varieties bringing a green revolution.

12.5 WIDE HYBRIDIZATION IN SOYBEAN

12.5.1 Selection of Parents in Wide Hybridization

- Figure 12.3 shows the geographical distribution of the genus *Glycine*. *Glycine soja*, the wild annual progenitor of *G. max* is indigenous to China, Korea, and Japan. By contrast, 26 wild perennial species of

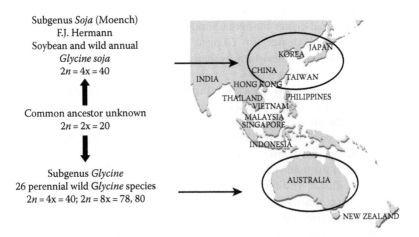

FIGURE 12.3 Geographical distribution of the *Glycine* species showing origin from a putative common ancestor still unknown.

the subgenus *Glycine* Willd., classified currently, are indigenous to Australia and adjoining islands. Thus, species of both subgenera are geographically isolated. *Glycine tomentella* is highly diverse morphologically, cytologically ($2n$ = 38, 40, 78, 80), has wide geographical distribution and harbors useful traits for resistance to pests and pathogens, and tolerance to abiotic stresses.

- The successful story of the production of modified soybean from *Glycine tomentella* (PI 441001; $2n$ = 78) as paternal and maternal parent with *Glycine max* cv. Dwight is described below.

  Story of *Glycine tomentella* (PI 441001; $2n$ = 78): This accession was collected from the Brampton Island off Queensland, Australia. *Professor A.H.D. Brown (personal communication 5/19/2015) described this accession as:* "There was something about this plant. It was collected by Paul Broué alone on a family vacation, as a good herbarium specimen with plenty of seed on the actual specimen. Paul died of a heart attack in 1983. The actual specimen was I believe the source of the seed that Ted introduced as IL428 (?) and from there into the PI system. Jim Grace regenerated seed from the specimen and gave it the number G1133, and very much by chance we included it as a standard on all our tomentella isozyme gels. I know it has a unique PGI isozyme pattern. Also by complete chance, whenever over the years we had requests for tomentella seed, G1133 was always included and this is how it got into the Monsanto program. Another incredible fact is that this sample of Paul's was our very first germplasm sample of Glycine tomentella." PI 441001 is resistant to soybean rust, soybean cyst nematode, and Phytophthora root rot.

- PI 441001 flowers profusely in a greenhouse (Figure 12.4a), contains a pair of nucleolus organizer chromosomes (Figure 12.4a insert bottom left); purple flowers are large and arranged on a long raceme (Figure 12.4b), produce black pods (Figure 12.4c insert top left) and small black seeds (Figure 12.4c).

- Soybean cultivar Dwight is annual (Figure 12.5a), contains a pair of nucleolus organizer chromosomes (Figure 12.5a insert bottom left) produces axillary purple flowers (Figure 12.5b); color of pods is tan with mostly one to three seeds per pod but occasionally four seeds per pod are produced (Figure 12.5c). Compared to seeds of PI 441001, Dwight seeds are yellow with black hilum (Figure 12.5d).

- Thus, morphologically, both parents are distinct.

12.5.2 Emasculation

- Selection of precise stage of soybean flower bud avoids self-pollination (Figure 12.6a). Remove calyx and corolla (Figure 12.6b).
- Remove (emasculate) all the 10 anthers carefully (Figure 12.6c).
- Dust pollen grains from a freshly bloomed flower (Figure 12.6d).

(a)   (c)

FIGURE 12.4 Morphological features of *Glycine tomentella;* PI 441001 (2n = 78) with a pair of nucleolus organizer chromosome (insert bottom left). (a) A plant at flowering stage with long narrow trifoliolate leaves, (b) a raceme with large purple flowers, and (c) black pods with 1–7 seeds (insert top left) and small black seeds.

(a)   (d)

FIGURE 12.5 (a) *Glycine max* cv. 'Dwight' with large trifoliolate leaves and maturing pod at each leaf node, contains a pair of nucleolus organizer chromosome; (b) Axillary purple color flower; (c) mature pods containing one to four seeds; (d) Mature, yellow, and large seeds with black hilum.

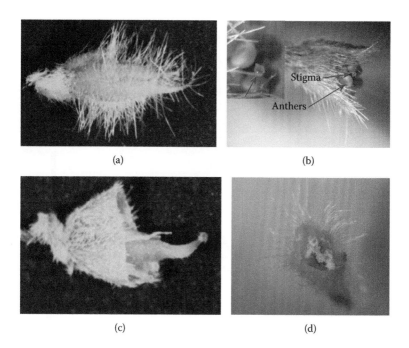

FIGURE 12.6 (a) The correct stage of flower bud of soybean cv. Dwight for emasculation; (b) flower bud without calyx and corolla showing sticky stigma (arrow) surrounded by anthers; (c) flower bud without calyx, corolla, and anthers; (d) Stigma covered with pollen grains of PI 441001.

12.5.3 EMBRYO RESCUE

- Composition of growth hormone
  - Gibberellic acid (GA₃): 100 mg
  - Naphthalene acetic acid (NAA): 25 mg
  - Kinetin: 5 mg
  - Distilled water: 1,000 mL
- Spraying of growth hormone
  - Spray growth hormone 24 h postpollination and continue once a day (morning) for 19–21 days.
  - Remove green pods from plant (Figure 12.7a).
- Immature seed rescue
  - Surface sterilize pods in a 3% solution of sodium hypochlorite (commercial bleach) for 20–30 min in a sanitized laminar flow cabinet.
  - Wash pods twice in sterile double-distilled water to remove all of the sodium chlorite.
  - Sterilize scalpel and forceps in Keller Steri 250 heating block.
  - Make a clean incision on the pod opposite to the funicle.
  - Remove immature green seeds and embed in the SM medium (Table 12.1).

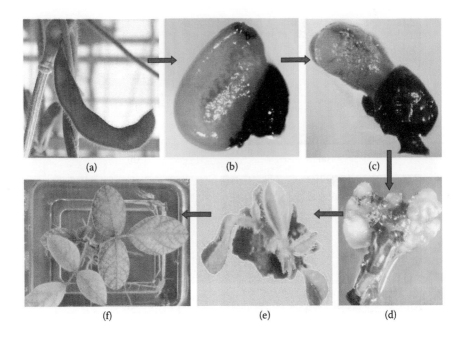

FIGURE 12.7 (a) One 21-days old hybrid pod showing three developing small seeds; (b) one immature seed in seed maturation (SM) medium after 1 week; (c) cotyledons emerging from a seed with black seed coat; (d) multiple embryos; (e) one developing shoot with leaves; (f) a small seedling with roots and shoot in a rotting medium.

TABLE 12.1

Alien Chromosome Substitutions and Transfer of Useful Traits to Cultigen from their Allied Species and Genera

Recipient	Donor	Procedure
Beta vulgaris	*Beta patellaris*	Substitution
Beta vulgaris	*Beta procumbens*	Substitution
Brassica napus	*Brassica nigra*	Substitution
Brassica napus	*Sinapsis arvensis*	Substitution
Cicer arietinum	*Cicer reticulatum*	Homoeologous pairing
Hordeum vulgare	*Hordeum bulbosum*	Irradiation-induced
Lycopersicon esculentum	*Lycopersicon chilense*	Homoeologous pairing
Lycopersicon esculentum	*Lycopersicon pennellii*	Homoeologous pairing
Lycopersicon esculentum	*Lycopersicon pennellii*	Homoeologous pairing
Lycopersicon esculentum	*Lycopersicon pennellii*	Homoeologous pairing
Lycopersicon esculentum	*Lycopersicon peruvianum*	Homoeologous pairing
Lycopersicon esculentum	*Solanum lycopersicoides*	Homoeologous pairing
Oryza sativa	*Oryza australeinsis*	Homoeologous pairing
Oryza sativa	*Oryza australeinsis*	Homoeologous pairing

(Continued)

TABLE 12.1 *(Continued)*
Alien Chromosome Substitutions and Transfer of Useful Traits to Cultigen from their Allied Species and Genera

Recipient	Donor	Procedure
Oryza sativa	*Oryza brachyantha*	Homoeologous pairing
Oryza sativa	*Oryza latifolia*	Homoeologous pairing
Oryza sativa	*Oryza latifolia*	Homoeologous pairing
Oryza sativa	*Oryza latifolia*	Homoeologous pairing
Oryza sativa	*Oryza longistaminata*	Homoeologous pairing
Oryza sativa	*Oryza minuta*	Homoeologous pairing
Oryza sativa	*Oryza minuta*	Homoeologous pairing
Oryza sativa	*Oryza minuta*	Homoeologous pairing
Oryza sativa	*Oryza nivara*	Homoeologous pairing
Oryza sativa	*Oryza officinalis*	Homoeologous pairing
Oryza sativa	*Oryza officinalis*	Homoeologous pairing
Oryza sativa	*Oryza officinalis*	Homoeologous pairing
Oryza sativa	*Oryza perennis*	Homoeologous pairing
Triticum aestivum	*Secale cereale*	1B(1R) Substituions and translocations
Triticum aestivum	*Secale cereale*	1B(1R) Substituions and translocations
Triticum aestivum	*Secale cereale*	1B(1R) Substituions and translocations
Triticum aestivum	*Secale cereale*	1B(1R) Substituions and translocations
Triticum aestivum	*Secale cereale*	1A.1R translocation
Triticum aestivum	*Secale cereale*	1A.1R translocation
Triticum aestivum	*Secale cereale*	1A.1R translocation
Triticum aestivum	*Secale cereale*	4A.2R translocation*4A.2R translocation*4A.2R translocation*4A.2R translocation*
Triticum aestivum	*Secale cereale*	4A.2R translocation*4A.2R translocation*4A.2R translocation*4A.2R translocation*
Triticum aestivum	*Secale cereale*	3A.3R translocation
Triticum aestivum	*Secale cereale*	3A.3R translocation
Triticum aestivum	*Secale cereale*	6BS.6RL translocation
Triticum aestivum	*Secale cereale*	2BS.2RL translocation
Triticum aestivum	*Secale cereale*	2BS.2RL translocation
Triticum aestivum	*Secale cereale*	1BL.1RS translocations
Triticum aestivum	*Secale cereale*	Irradiation induced translocations
Triticum aestivum	*Secale cereale*	4BL.5RL translocation
Triticum aestivum	*Secale cereale*	2A.2R translocation
Triticum aestivum	*Triticum monococcum*	Bridge cross
Triticum aestivum	*Agropyron elongatum*	Irradiation induced translocation
Triticum aestivum	*Agropyron elongatum*	Irradiation induced translocation
Triticum aestivum	*Agropyron elongatum*	Homoeologous pairing
Triticum aestivum	*Agropyron elongatum*	Substitution
Triticum aestivum	*Agropyron elongatum*	Substitution

(Continued)

TABLE 12.1 *(Continued)*
Alien Chromosome Substitutions and Transfer of Useful Traits to Cultigen from their Allied Species and Genera

Recipient	Donor	Procedure
Triticum aestivum	*Agropyron intermedium*	Translocations
Triticum aestivum	*Agropyron intermedium*	Irradiation induced translocation
Triticum aestivum	*Aegilops comosa*	Homoeologous pairing
Triticum aestivum	*Aegilops longissimum*	Homoeologous pairing
Triticum aestivum	*Aegilops sharonensis*	Irradiation induced
Triticum aestivum	*Aegilops speltoides*	Homoeologous pairing
Triticum aestivum	*Aegilops umbellulata*	Irradiation induced translocation
Triticum aestivum	*Aegilops variabilis*	Homoeologous pairing
Triticum aestivum	*Aegilops ventricosa*	Homoeologous pairing
Triticum aestivum	*Aegilops uniaristata*	Homoeologous pairing
Triticum aestivum	*Thinopyrum distichum*	Translocaiton
Triticum aestivum	*Thinopyrum intermedium*	Translocation
Triticum aestivum	*Thinopyrum intermedium*	Addition
Triticum aestivum	*Thinopyrum intermedium*	Addition
Triticum aestivum	*Thinopyrum intermedium*	Substitution
Triticum aestivum	*Triticum araraticum*	Homoeologous pairing
Triticum aestivum	*Triticum monococcum*	Homoeologous pairing
Triticum aestivum	*Triticum turgidum*	Homoeologous pairing
Triticum aestivum	*Haynaldia vilosa*	Substitution
Triticum aestivum	*Agropyron intermedium x Triticum monococcum*	Substitutions and translocations
Triticum aestivum	*Aegilops speltoides x Triticum monococcum*	Homoeologous pairing
Avena sativa	*Avena barbata*	Irradiation -induced translocation
Avena sativa	*Avena abyssinica x Avena strigosa*	Irradiation induced translocations
Trifolium repens	*Trifolium nigrescens*	Polyploidization

Source: Singh, R.J., *Plant Cytogenetics*, 3rd ed., CRC Press, Boca Raton, FL, 2017.
Note: Agropyron elongatum, Elytrigia elongata, Thinopyrum elongata, and Lophopyrum elongatum are one species.

- Seal Petri plates with a strip of Parafilm (Bemis Flexible Packaging—Neenah, WI 54956).
- Keep Petri plates with seeds in an incubator with 13 h of low-intensity light (25–30 μmol photons/m^2/s) at 25°C.
- Transfer seeds every 3–4 weeks to fresh SM medium (Figure 12.7b).
- Continued this process until cotyledons (Figure 12.7c) emerge.

- Transfer germinating seeds to C5 medium (Table 1) and after few transfers embryogenic calluses emerge (Figure 12.7d).
- After few transfers to C5 medium, multiple shoots appear (Figure 12.7e).
- Transfer actively growing shoots to rooting medium (Table 1). Roots develop within 3–4 weeks (Figure 12.7f).

12.5.4 PRODUCTION OF F_1 AND AMPHIDIPLOID PLANTS

- Transfer seedlings in pots with soil in the greenhouse, cover seedlings with a plastic bag.
- Remove plastic bag slowly to acclimatize the seedlings to greenhouse growing conditions.
- Seedling growth is slow initially but once the root system is developed, growth is normal (Figure 12.8a).
- To verify the hybrid nature, collect roots and examine chromosomes at mitotic metaphase and that F_1 plants contain expected $2n = 59$ chromosomes (Figure 12.8b).
- F_1 plant is totally sterile because of abnormal meiosis (Figure 12. 8c).
- Treat actively growing shoots in a filter sterilized solution of 0.2% colchicine + 2% dimethyl sulfoxide (DMSO) + 10 drops Tween® 20 + 10 mg GA3/L for 4–6 h in the culture incubator.
- Wash shoots twice in sterile distilled water in a sanitized laminar flow cabinet, and transfer shoots to C5 medium.

(a) (b) (c)

FIGURE 12.8 (a) F_1 plant growing in pot with soil in greenhouse; (b) mitotic metaphase chromosome of F_1 plant (two satellite chromosome, arrows); (c) meiotic metaphase cell showing mostly univalents and 5 rod-shaped bivalents.

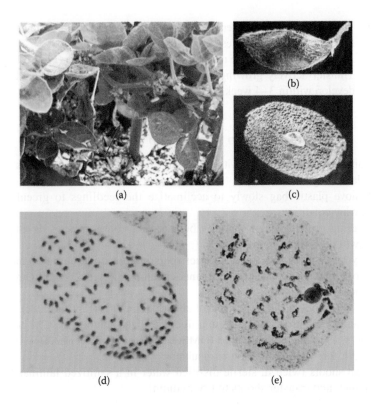

(a)   (b)   (c)   (d)   (e)

FIGURE 12.9   (a) Amphidiploid plant showing stunted vegetative growth, and developing pod (arrow); (b) one-seeded mature pod; (c) a mature seed; (d) a mitotic metaphase cell with $2n = 118$; and (e) diakinesis cell showing mostly bivalents configuration.

- Amphidiploid is slow in growth with dark green leaves (Figure 12.9a) and occasionally produces 1–2 seeded mature pods (Figure 12.9b) and seeds (Figure 12.9c).
- Examine chromosome at mitotic metaphase to confirm the validity of amphidiploid; $2n = 118$ (Figure 12.9d). During diakinesis, mostly bivalents (Figure 12.9e) are observed.

12.5.5   PRODUCTION OF BC_1 AND BC_2 PLANTS

- Hybridize amphidiploid plant by Dwight (recurrent parent) pollen.
- Culture immature seeds (Figure 12.10a) in SM medium.
- Induce multiple shoots (Figure 12.10b).
- Transfer BC_1 plants in pots with soil in greenhouse.
- Identify plants morphologically (Figure 12.10c) and verify cytologically.
- BC_1 plants contained expected $2n = 79$ chromosome (40 chromosomes from Dwight + 39 chromosomes from PI 441001).
- Pollinate BC_1 plants with Dwight. Mature BC_2F_1 seeds were obtained.

FIGURE 12.10 (a) BC_1 seed derived from amphidiploid x Dwight; (b) multiple shoots; (c) a BC_1 plant growing in the greenhouse showing twinning trait of PI 441001; (d) 06H1-1 ($2n = 58$), BC_2F_1 plant, a black pod with two seeds (insert right bottom), and greenish grey seed (insert right top); (e) 06H1-3 ($2n = 56$), BC_2F_1 plant, produced yellow with black hilum seed (insert right top); (f) 07H1-7 ($2n = 56$), BC_2F_1 plant, produced black pods and black seeds (insert right top); (g) 07H6-3, BC_3F_1 plant ($2n = 41$) produced black pods with grey seeds; (h) 07H5-8, BC_3F_1 plant ($2n = 41$) produced normal seeds like recurrent parent; (i, j) Arrow shows *Glycine tomentella* chromosome; (k) a metaphase cell stained through GISH showing on PI 441001 chromosomes; and (l) derived lines growing in the field (2008).

- Of the 3006 BC_1 flowers pollinated, 30 seeds germinated BC_2F_1 plants segregated for 2n = 55–99 chromosomes.
- Only 15 BC_2F_1 plants survived past flowering and these plants contained $2n = 56$ (nine plants), $2n = 57$ (two plants), $2n = 58$ (three plants), and $2n = 59$ (one plant).
- Figure 12.10d shows a plant with $2n = 58$ (06H1-1), which grew like a small shrub, was vigorous, and flowered profusely. It produced 43 BC_3F_1 seeds from 1063 flowers pollinated with Dwight. Pods turned black (Figure 12.10d, right bottom insert) contained one to two seeds which had greenish-brown seed coats (Figure 12.10d, right top insert).
- Plant 06H1-3 ($2n = 56$; Figure 12.10e) grew slowly—it was only a few centimeters tall. It produced small, narrow leaves with short petioles, and flowered earlier than the recurrent parent. Most of the flowers were cleistogamous. It produced 43 seeds from 466 flowers pollinated. Pod color was black but all seeds were round, yellow, with dark black hilum. Seed size was smaller (Figure 12.10e, insert right top) than the recurrent parent.

- Plant 07H1-26 ($2n = 56$) was bushy; leaves were long, narrow, and droopy (Figure 12.9f), and it produced black pods and black seeds (Figure 12.9f insert right top).

12.5.6 Production of BC$_3$F$_1$ Plants

- BC$_3$F$_1$ of 06H1-1 segregated plants with $2n = 41$–49, 06H1-3 segregated plants for $2n = 41$–49 and 07H1-26 segregated plants with $2n = 42$–49.
- BC$_3$F$_1$ plant 07H6-3 ($2n = 41$) was slow in vegetative growth (Figure 12.10g) compared to Dwight, produced black pods and gray seeds (Figure 12.10h).
- BC$_3$F$_1$ plant 07H5-8 ($2n = 41$) resembled Dwight, highly fertile (Figure 12.10i) and GISH showed one PI 441001 labeled chromosome (Figure 12.9j; arrow). This plant produced seeds (Figure 12.10k) like Dwight.

12.5.7 Production of BC$_3$F$_2$ Plants and Beyond

BC$_3$F$_2$ plants growing in the field during 2008 (Figure 12.10).

- Hybridization started in September 1, 2003 and fertile plants were moved to the field in 2008, after 5 years.
- Materials are still (2016) being produced from the BC$_3$F$_1$, BC$_4$F$_1$, and BC$_5$F$_1$ plants with higher chromosome numbers.

12.5.8 Isolation of Monosomic Alien Addition Lines

- Figure 12.11a shows the methodology to isolate monosomic alien addition lines (MAALs) from Dwight cytoplasm.
- Figure 12.11b shows the methodology to isolate MAALs from PI 441001 cytoplasm.
- In *Glycine max* cv. Dwight ($2n = 40$; G$_1$G$_1$) x *Glycine tomentella* ($2n = 78$; D$_3$D$_3$EE), it is expected to isolate 39 different types of MAALs.
- BC$_2$F$_1$ plants segregate for plants with $2n = 55$–59 (G$_1$G$_1$ + D$_3$ + E).
- BC$_3$F$_1$ plants segregate for chromosomes $2n = 40$–49.
- During BC$_4$-BC$_6$; F2$_{-4}$ generations, identify plants cytologically with $2n = 40$ and 41 (Singh and Nelson 2015).

12.5.9 Broadening the Cytoplasm of the Soybean

- In order to diversify the cytoplasm of soybeans, hybridize *Glycine tomentella*, PI 441001 ($2n = 78$) as a female parent.
- Follow the protocols described for *Glycine max* cv. Dwight × *Glycine tomentella*, PI 441001 (Figure 12.12).
- Meiotic chromosome pairing at diakinesis showed one quadrivalent (Figure 12.13a) and bivalent associations. At metaphase-I, bivalents were

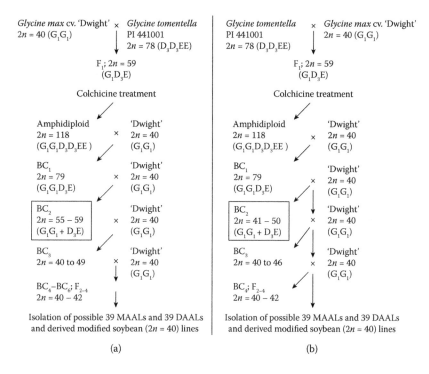

FIGURE 12.11 Diagrammatic sketch showing isolation of monosomics and disomic alien addition lines from *Glycine max x G. tomentella and G. tomentella x G. max.*

arranged at the equatorial plate with ring- and rod-shaped bivalents while univalent precociously moved to the poles.

- Clear bivalent configurations (end-to-end synapsis) in F_1 sporocytes during pachynema (Figure 12.13c through e) suggest that gene exchange between Dwight and PI 441001 is possible.
- Amphidiploid plants were vigorous and flowered profusely (Figure 12.14a). The true nature of amphidiploid was verified by a mitotic metaphase chromosome which showed $2n=118$ chromosomes (Figure 12.14b).
- Amphidiploid × Dwight produced mature back pods and seeds (Figure 12.14c through d). This was not observed in a Dwight similar cross in Dwight cytoplasm.
- Morphologically, BC_1 plants were weak, slow growing with curved (inward) dark green and thick leaves. They produced one to two seeded pods (Figure 12.14f and g). Seeds were round, small, yellow-gray with a black hilum (Figure 12.14h).
- Chromosome segregation in BC_2F_1 ranged from $2n = 41–50$ (Figure 12.11b) which was observed in BC_3F_1 in Dwight cytoplasm (Figure 12.11a).
- In BC_2F_1, four plants with $2n = 44$ chromosomes were identified but were morphologically different (Figure 12.15a through d).
- Figure 12.16 shows $2n = 40$ (a) and four (b, c, d, and e) MAALs.

FIGURE 12.12 Production of fertile plants from *Glycine tomentella* PI 441001 ($2n = 78$) x *G. max* cv. Dwight ($2n = 40$). (a) immature 21-day-old seed in culture, (b) embryogenic callus extruded leaving behind black seed coat, (c) multiple shoot generation from an embryogenic callus, (d) the F_1 hybrid plant growing in greenhouse, and (e) mitotic metaphase cell with $2n = 59$ chromosomes from the F_1 plant.

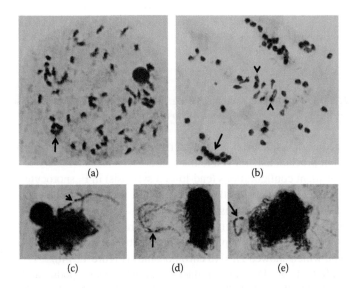

FIGURE 12.13 Meiosis in F_1 plant from *Glycine tomentella* PI 441001 ($2n = 78$) x *G. max* cv. Dwight ($2n = 40$). (a) Diakinesis cell showing a quadrivalent (ring of 4; arrow), loosely associated rod-shaped bivalents and univalents; (b) metaphase I cell showing five rod-shaped bivalents (chiasma terminalized), and two ring-shaped bivalents (arrow heads). Movement of univalents was precocious and reached the telophase poles, and few univalents remained at the equatorial plate. Arrow shows an aggregation of six univalents and can be misidentified as a hexavalent; (c) pachynema cell showing a bivalent synapsed end-to-end with differentiated euchromatin and heterochromatin (arrow shows kinetochore); (d) pachynema cell showing a bivalent synapsed end-to-end and only kinetochore is flanked by heterochromatin (arrow); (e) pachynema cell showing a bivalent synapsed end-to-end; this bivalent is the smallest chromosomes (chromosome 20 of Singh and Hymowitz, 1988) short arm is heterochromatic and half of the long arm is heterochromatic.

FIGURE 12.14 Amphidiploid and BC_1 plants of *Glycine tomentella* PI 441001 ($2n = 78$) x *G. max* cv. Dwight ($2n = 40$). (a) Amphidiploid plant growing in greenhouse; plant has twinning habit like PI 441001; (b) mitotic metaphase cell from Figure 3a plant showing $2n = 118$ chromosomes; (c) two mature BC_1 pods each containing one seed; (d) mature BC_1 seed showing black seed coat like PI 441001; (e) BC_1 plant with $2n = 79$ chromosomes growing in the greenhouse. Plant is not a climber and has dark-green thick curved leaves; (f) green BC_1F_1 pods developing at the base of stem; (g) two cross pods at a node of BC_1 plant: one pod (tan) is close to maturity and second pod is still green; (h) mature BC_2F_1 seed showing tan seed coat and black hilum.

- It is easy to identify MAAL with three satellite chromosomes (Figure 12.17a).
- Meiotic metaphase chromosome pairing usually shows 20II + 1I (Figure 12.17b).
- Plants with 40 chromosomes occasionally expressed total sterility (Singh and Nelson 2014).

(a)   (b)   (c)   (d)

FIGURE 12.15 Four BC$_2$F$_1$ plants with $2n = 44$ chromosomes showing differences among plants due to the combination of PI 441001 chromosomes. (a) Plant 10H270-1 showing dwarf growth habit, dark-green curved leaves, flowers on node (few) and pods hidden under the leaves, mostly one-seeded pod (insert); (b) plant 10H270-5 showing yellowish leaves and pods (insert); (c) plant 10H270-9 showing dark-green leaves with normal pod set like Dwight; and (d) plant 10H270-10 showing slow vegetative growth with dark-green narrow leaves compared to other three fertile 44-chromosome plants (a–c).

(a)   (b)   (c)   (d)   (e)

FIGURE 12.16 Isolation and identification of four morphologically distinct MAALs ($2n = 41$) derived from nine seeds of a BC$_3$F$_1$ 13ST500; 11H1-4, $2n = 47$ x Dwight. (a) Plant 13ST500-1 with $2n = 40$ and has mature pods with green leaves; (b) Plant 13ST500-2 with $2n = 41$ with normal fertility but irregular pods at maturity; (c) plant 13ST500-3 plant with $2n = 41$ showing one main branch (arrow), woody, set few flat one-seeded pods; (d) plant 13ST500-8 with $2n = 41$, plant was slow in vegetative growth with dark-green thick curved leaves, produced few self-large pods. (e) Plant13ST500-9 plant with $2n = 41$ showing droopy trifoliolate leaves with long petiole (arrow) and poor pod set compared to (a) and (b).

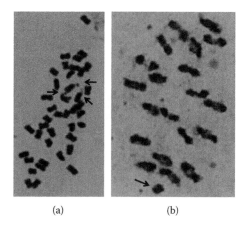

(a)   (b)

FIGURE 12.17 Mitotic metaphase and meiotic metaphase I of plant 13ST500-8; (a) mitotic metaphase cell with $2n = 41$ chromosome showing three satellite chromosomes (arrows); (b) meiotic metaphase I microsporocyte of plant 13ST500-8 showing 20II (Dwight chromosomes) + 1I (PI 441001 chromosome)

12.6 INTROGRESSION OF USEFUL GENES THROUGH DISTANT HYBRIDIZATION

- Distant hybridization has played a key role in transferring useful genes such as resistance to pests and pathogens from alien species (GP-2 and GP-3) to the cultigens, particularly in wheat and rice.
- Table 12.1 summarizes the transfer of useful genes in many crops from distantly related species (Singh 2017).

REFERENCES

Harlan, J. R., and J. M. J. de Wet. 1971. Toward a rational classification of cultivated plants. *Taxon* 20: 509–517.

Singh, R. J. 2017. *Plant Cytogenetics*. 3rd edn. CRC Press, Boca Raton, FL.

Singh, R. J., and R. L. Nelson. 2014. Methodology for creating alloplasmic soybean lines by using *Glycine tomentella* as a maternal parent. *Plant Breed.* 133: 624–631.

Singh, R. J., and R. L. Nelson. 2015. Intersubgeneric hybridization between *Glycine max* and *G. tomentella*: Production of F_1, amphidiploid, BC_1, BC_2, BC_3, and fertile soybean plants. *Theor. Appl. Genet.* 128: 1117–1136.

Glossary

ABA: Abscisic acid.

Aberration: Variation of chromosome structure either induced by mutagen or natural mutation.

Aberration rate: The portion of chromosomal changes as compared to normal chromosomes.

Abort: The collapse of seeds prior to maturation.

Abortive: Defective or barren; development arrested while incomplete or imperfect.

Abscise: Separation of leaves from stem or flowers from raceme.

Abscission: Rejection of plant organs.

Abscission layer: Cells of abscission zone disintegrate causing the separation of plant organs (leaf, fruit, pod, and branch).

Abscission zone: Layer of cells which brings about the abscission of plant organs.

Abscisic acid: Plant growth hormone involved in fruit growth and rejection of plant organs.

Accessions: A sample of plant germplasm maintained in a gene bank for preservation and utilization by scientists.

Accession number: Each accession is assigned a particular number; such as PI (plant introduction) number in the USDA gene bank.

Accessory chromosome: Supernumerary or B-chromosome.

Acentric fragment: A chromosome fragment without a kinetochore.

Acetocarmine stain: A dye (prepared in 45% acetic acid + 1%–2% carmine powder) used for staining plant chromosomes.

Aceto-orcein stain: In this stain, carmine is replaced by orcein.

Achiasmate: Meiosis without chiasma. Chromosome 4 in *Drosophila* which is mostly heterochromatic is without chiasma. Heterochromatin has been implicated in chromosome pairing for achiasmatic disjunction.

Achromatic: Chromosome regions which do not stain with specific dyes.

Achromatin: The part of the nucleus that does not stain with basic dyes.

A-chromosome: One of the standard chromosomes of an organism.

Acquired characters: Changes in phenotype of an organism arise purely by environmental influences during development and not the result of gene action (Lamarckian inheritance).

Acquired immunity: Resistance developed in an organism during preimmunization.

Acquired mutation: A noninheritable genetic change in an organism—somatic mutation.

Acridine orange: A dye that functions as a fluorochrome and a mutagen.

Acrocentric chromosome: A chromosome consisting of one complete arm and a small piece of the other arm.

Acrosyndesis: Incomplete end to end chromosome pairing.

Acrotrisomics: An individual with a normal chromosome complement plus an extra acrocentric chromosome.

Addition line: A cell line of individual carrying chromosomes or chromosome arms in addition to the normal standard chromosome set.

Adjacent-1 distribution: In an interchange, heterozygote nonhomologous kinetochores move to opposite poles.

Adjacent-2 distribution: In an interchange, heterozygote homologous chromosomes move to opposite poles.

Adjacent segregation: In reciprocal translocation heterozygote, the segregation of translocated and normal chromosomes gives unbalanced spores with duplications and deficiencies.

Adosculation: The fertilization of plants by pollen falling on the pistils.

Adventitious embryony (Adventive embryony): Embryo develops from the nucellus or integument cells of the ovule.

AFLP: Amplified fragment length polymorphisms.

Albuminous seed: A mature seed which contains endosperm.

Alien addition line: A line containing an unaltered chromosome complement of one species (generally a cultigen) plus a single chromosome (monosomic alien addition lines, MAALs) or a pair of chromosomes (disomic alien addition lines, DAALs) from an alien species.

Alien chromosome: A chromosome from a related species.

Alien chromosome transfer: Transfer of individual alien chromosomes from alien species to cultigen through cytogenetic methods.

Alien gene transfer: The transfer of genes from exotic species to cultigen.

Alien germplasm: Genes from distantly wild species.

Alien species: An organism that has invaded or been introduced by man and is growing in a new region; a species that serves as a donor of chromosomes, or chromosome segments to be transferred to a recipient species or genotype.

Alien substitution line: A line in which cultivated one or more chromosomes are substituted from one or more chromosomes of an alien species.

Allele: An allele is merely a particular form of the same gene, can be dominant or recessive, and can be in homozygous and heterozygous conditions.

Allele mining: An approach to access new and useful genetic variation in crop plant collections.

Allelic: Two genes similar—same gene.

Allelism: The genes occupying the same relative position (locus) on homologous chromosomes.

Allelopathic: The populations are not in physical contact and live in different geographic areas.

Allodiploid: Cells or individuals in which one or more chromosome pairs are exchanged for one or more pairs from another species.

Allogamy: Cross-fertilization.

Allogene: Recessive allele.

Allogenetic: Cells or tissues related but sufficiently dissimilar in genotype to interact antigenically.

Allohaploid: A haploid cell or individual derived from an allopolyploid and composed of two or more different chromosome sets.

Alloheteroploid: Heteroploid individuals whose chromosomes derived from various genomes.

Allopathy: Different populations are not geographically in contact.

Alloplasm: Cytoplasm from an alien species that has been transferred by backcrossing into a cultivated species.

Alloplasmic: An individual having cytoplasm of an alien species and nucleus of a cultivated species.

Allopolyploid: An organism containing different sets of chromosomes derived from two or more genomically diverse species. For example, allotetraploid, allohexaploid, and alloctoploid.

Allosyndesis: Meiotic chromosome pairing of completely or partially homologous (homoeologous) chromosomes in allopolyploids.

Allotetraploid: A tetraploid individual having two different diploid genomes.

Alternate segregation: In reciprocal translocation heterozygote, during meiotic anaphase-I, the movement of both normal chromosomes to one pole and both translocated chromosomes to the other pole, resulting in genetically balanced spores.

Alternation of generation: In the life cycle of many plants, a gametophyte phase alternates with a sporophyte phase.

Alternative disjunction: The distribution of interchange chromosomes at anaphase-I is determined by their centromere orientation; chromosomes located alternatively in the pairing configuration are distributed to the same spindle pole.

Ameiosis: The failure of meiosis and its replacement by nuclear division without reduction of the chromosome number.

Ameiotic mutants: Meiosis does not ensue and pollen mother cells degenerate.

Amitosis: Absence of mitosis.

Amixis: Reproduction in which the essential events of sexual reproduction are absent.

Amphicarpous: A plant producing two classes of fruit that differ either in form or in time of ripening.

Amphidiploid: An individual which contains entire somatic chromosome sets of two species produced after chromosome doubling and exhibits diploid-like meiosis.

Amphimixis: Seed develops by a fusion of a female and a male gamete—normal sexual reproduction.

Amphitene: Zygonema or zygotene.

Amplification: The production of additional copies of a chromosome sequence, found as intrachromosomal or extrachromosomal DNA.

Anandrous: Free from anthers.

Anaphase (mitosis): Anaphase is the shortest stage of mitosis. Mutual attachment between homologous chromosomes is ceased, sister chromatids at the kinetochore region at metaphase repulse and two chromatids, equal distribution of chromosomes, move toward the opposite poles at the spindle.

Anaphase-I (meiosis): Migration of homologous chromosomes to the opposite poles.

Androecium: All stamens of a flower; the male reproductive organ of a plant.

Androgenesis: Synonymous: male parthenogenesis. The progeny inherits the genotype of the male gamete nucleus.

Androsome: Any chromosome exclusively present in the male nucleus.

Aneucentric: Chromosome possessing more than one centromere.

Aneuhaploid: Chromosome number in aneuhaploid is not an exact multiple of the basic set but an individual is deficient for one or more chromosomes.

Aneuploid: An organism or species containing chromosomes other than an exact multiple of the basic (x) chromosome number.

Aneuploid reduction: Reduction of the genetic variability by decreasing the chromosome number.

Aneusomatic: An organism having both euploid and aneuploid cells.

Angiosperm: Seeds formed within an ovary.

Anisoploid: An individual with an odd number of chromosome sets in somatic cells.

Annealing: The pairing of complementary single strands of DNA to form a double helix.

Anther: Part of the stamen bearing pollen grains.

Anthesis: The time of flowering in a plant.

Antimitotic: Substance that leads to the cessation of mitosis.

Antipodal cells: A group of three cells each with a haploid nucleus formed during postmegasporogenesis, located at the chalazal end of the mature embryo sac.

Anucleolate: Without nucleus.

Apogamety: Embryo is formed from synergids and antipodals other than the egg without fertilization.

Apomeiosis: A type of gametophytic apomixis where failure of both chromosome reduction and recombination occurs.

Apomict: An organism produced by apomixis.

Apomixis: Apomixis is a mode of asexual reproduction in plants where seed is produced without sexual fusion of female and male gametes. Synonymous: agamospermy.

Apospory: Embryo sac is formed from somatic cells in the nucellus or chalaza of the ovule.

Asexual reproduction: A type of reproduction that does not involve the union of male and female gametes; vegetative propagation.

Asymmetric cell division: Division of a cell into two cells of different chromosome numbers.

Asymmetrical fusion: A cell formed by the fusion of dissimilar cells; heterokaryon.

Asymmetrical karyotype: Characterized by the set containing some large and some small chromosomes.

Asynapsis: A complete failure of homologous chromosome pairing during meiosis.

Autobivalent: A bivalent at meiotic metaphase-I that is formed from two structurally and genomically completely identical chromosomes.

Auto-fertility: Self-fertility.

Autogamy: Self-pollination predominates; this is a common and widespread condition in angiosperms, particularly in annual herbs.

Autogenomatic: Genomes that are completely homologous and pair normally during meiosis.

Automixis: Fusion of two haploid nuclei in a meiotic embryo sac without fertilization of the two egg nucleus, producing maternal-like progeny.

Autoradiography: This method detects radioactivity labeled molecules by their effects in creating an image on photographic film.

Autosome: Any chromosome other than sex (X or Y) chromosome.

Autosyndesis: The pairing of complete or partial homologues of chromosome.

Autotetraploid (4x, AAAA): An individual possessing four identical sets of homologous chromosomes.

Autotriploid (3x, AAA): An individual possesses three identical sets of homologous chromosomes.

Autozygote: A diploid individual in which the two genes of a locus are identical by descent.

Auxin: Hormones that promote cell elongation, callus formation, fruit production, and other functions in plants.

B–A translocation: These interchanges are obtained by breakage in the A chromosomes and in the B-chromosomes and after reciprocal exchanges, A^B and B^A chromosomes are produced.

bp: An abbreviation of base pairs; distance along DNA is measured in bp.

Backcross: The crossing of an F_1 heterozygote to one of its parents.

Balanced lethal: A situation in which only certain combinations of gametes survive in zygotes. For example, heterozygous velans/gaudens survive and homozygotes die in *Oenothera lamarckiana*.

Balbiani ring: An extremely large puff at a band of a polytene chromosome.

Banding pattern: The linear pattern of deeply and weakly stained bands on the chromosomes after C-N-G-banding techniques.

Base pairs: A partnership of A (adenine) with T (thymine) or of C (cytosine) with G (guanine) in a DNA double helix. In an RNA, thymine is replaced by uracil (A = U; C = G).

Base sequence: The order of nucleotide bases in a DNA molecule.

Base sequence analysis: A technique, often automated, for determining the base sequence.

Base substitution: A form of mutation in which one of the bases in DNA replaces another.

Basic chromosome number: The haploid chromosome number of the ancestral species represented as x (basic genome). For example, hexaploid wheat is identified as $2n = 6x = 42$ (AABBDD).

B-chromosome: These are also called supernumerary chromosomes, derived from the A chromosome complement; usually heterochromatic but genetically not completely inert.

Bimitosis: The simultaneous occurrence of two mitoses in binucleate cells.

Binucleate: Cells with two nuclei.

Biotechnology: The set of biotechnology methods developed by basic research and now applied to research (recombinant DNA technology) and product development.

Bipartitioning: Normal meiosis.

Bisexual (= perfect): A flower containing both stamens and carpels.

Bivalent: A pair of homologous chromosomes synapsed during meiosis.

Bivalent formation: The association of two homologous chromosomes as a ring or rod configuration depending on chiasma formation.

Botany: The science of biology that deals with plant life.

Breakage-fusion-bridge cycle: A process that can arise from the formation of dicentric chromosomes; daughter cells are formed that differ in their content of genetic material due to duplications and/or deletions in the chromosomes.

C-banding: Constitutive heterochromatin exhibits bands after staining with Giemsa, is redundant and usually present in the proximity of kinetochore, telomere, and in the nucleolus organizer region.

C-value: The total amount of DNA in a haploid genome.

Calyx: All the sepals (outer part) of a flower.

Carmine: A red dye used o stain the chromosomes; usually it is dissolved in 45 % acetic acid and known as aceto-carmine.

Carnoy's fixative: A fixative used for cereal chromosome study; consists of 6 parts 95% ethanol:3 parts chloroform:1 part acetic acid.

Carpel: Pistil; a female reproductive organ including ovary, style, and stigma.

Cell: The smallest, membrane-bound, living unit of biological structure capable of auto-reduplication.

Cell cycle: The period from one division to the next; consists of G_1, S, G_2, and M.

Cell division: The reproduction of a cell by karyokinesis and cytogenesis.

Cell furrow: The wall formed between the two-daughter nuclei at the end of telophase in animals.

Cell plate: The wall formed between the two-daughter nuclei after karyokinesis.

Centimorgan (cM): A unit of measure of recombination frequency.

Centric: A chromosome segment having kinetochore.

Centric fission: A chromosomal structural change that results in two acrocentric or telocentric chromosomes, as opposed to centric fusion.

Centriole: Small hollow cylinders consisting of microtubules that become located near the poles during mitosis.

Centromere: see kinetochore.

Centromere interference: An inhibitory influence by the centromere on crossing-over and the distribution of chiasmata in its vicinity.

Centromere misdivision: A transverse instead of lengthwise division of the centromere resulting in telocentric chromosomes.

Centromeric index: CI = short arm length / chromosome length × 100.

Centrosome: The self-propagating body which divides during cell division in many organisms to form the two poles of the spindle and determine its orientation.

Character: Phenotypic trait of an organism controlled by genes and may be dominant or recessive; may be governed by a single gene or multigenes.

Chiasma: An exchange of chromatid segment during the first division of meiosis that results in a genetic crossing-over; plural: chiasmata.

Chiasma interference: The occurrence, less frequent or more frequent than expected by chance, of two or more crossing-over and chiasmata in a given segment of a chromosomal pairing configuration and/or chromosome.

Chiasma localization: The physical position of a chiasma in a pairing configuration and/or chromosome.

Chiasmate: Meiosis with normal chiasma formation.

Chiasma terminalization: The progressive shift of chiasmata along the arms of paired chromosomes from their points of origin toward terminal positions.

Chimaera: A plant composed of two or more genetically distinct types of tissues due to mutation, segregation, and irregularity of mitosis (mosaic).

C-mitosis: Complete inactivation of spindles producing a restitution nucleus with double chromosome number by colchicine treatment.

Chromatid: A half-chromosome (sister chromatids) that originates from longitudinal splitting of a chromosome during mitotic anaphase or meiotic anaphase-II.

Chromatid aberration: Chromosomal changes produced in one chromatid due to result of spontaneous or induced mutation.

Chromatid Bridge: Dicentric chromatid with kinetochores passing to opposite poles at anaphase.

Chromatid segregation: Double reduction, used in autopolyploidy.

Chromatin: The complex of DNA and protein in the nucleus of the interphase cell; originally recognized by reaction with stains specific for DNA.

Chromatin reconstitution: The reconstitution of chromatin with chromosomal constituents previously removed by chromatin dissociation.

Chromocenter: Darkly stained bodies in the interphase nuclei consisting of fused heterochromatic telomeres.

Chromomere: Knob-like regions along the entire length of pachytene chromosomes occupy specific positions on the chromosomes, larger heterochromatic chromomeres stain relatively darker than the euchromatic chromomeres.

Chromomere pattern: The linear order and distribution of chromomeres along a chromosome.

Chromonema: The chromosome threads at leptonema.

Chromosomal: Relating to the structure, constituents, and function of chromosomes.

Chromosomal aberrations: An abnormal chromosomal complement resulting from the loss, duplication, and rearrangement of genetic materials.

Chromosomal inheritance: The inheritance of genetic traits located in the chromosomes (Mendelian inheritance).

Chromosome: An important, complex organelle of the nucleus, containing genes. The number is maintained constant from generation to generation by mitotic and meiotic divisions; it consists of protein, nucleic acid, and genes.

Chromosome arm: One of the two parts of a chromosome divided by an intercalary centromere.

Chromosome complement: The total chromosome constitution of an organism.

Chromosome configurations: An association of chromosomes during meiosis in the form of univalent, bivalent, trivalent, quadrivalent (tetravalent), and so on.

Chromosome congression: Chromosomes move to the equatorial plate halfway between two poles of the spindle with the help of spindle fibers.

Chromosome drift: The shift in the chromosome number.

Chromosome elimination: The loss of chromosomes from nuclei due to genic, chromosomal, or genomic causes.

Chromosome fragmentation: The breakage of chromosomes by radiation.

Chromosome map: A linear representation of a chromosome in which genetic markers belonging to a particular linkage group are plotted based on their relative distances—physical mapping. Chromosome maps are: (1) cytological maps—according to karyotype of chromosomes; (2) genetic maps—according to crossing-over, recombination, and map units; (3) cytogenetic maps—gene located in a particular chromosome based on cytologically identifiable chromosome aneuploids such as deletions, duplications, translocations, inversions, trisomics, and other chromosomal aberrations; (4) restriction fragment length polymorphism (RFLP) map—a linear graph of sites on DNA cleaved by various restriction enzymes.

Chromosome mosaicism: The presence of cell populations of various karyotypes in the same individual.

Chromosome movement: The movement of chromosomes during mitosis and meiosis as a prerequisite for the anaphase separation of chromatids and/or chromosomes.

Chromosome mutation: Any structural change involving the gain, loss, or translocation of chromosome parts; it can arise spontaneously or be induced experimentally by physical and chemical mutagens.

Chromosome number: The specific somatic chromosome number (2n) of a given species.

Chromosome pairing: The highly specific side-by-side association of homologous chromosomes during meiotic prophase.

Chromosome segregation: The separation of the members of a pair of homologous chromosomes in a manner that only one member is present in any postmeiotic nucleus.

Chromosome set: A minimum complement of chromosomes derived from the gametic complement of a putative ancestor (basic chromosome-genome).

Chromosome size: The physical shape and size of a chromosome.

Chromosome substitution: The replacement of one or more chromosomes by others from another source by spontaneous events or crossing-over.

Chromosome walking: This describes the sequential isolation of clones carrying overlapping sequences of DNA, allowing regions of the chromosome to be spanned. Walking is often conducted in order to reach a particular locus.

Cleavage: Division of cytoplasmic portion in animals.

Cleistogamous flower: Pollination and fertilization occur before flowers open ensuring complete self-pollination.

Clone: A group of organisms descended by mitosis, that is, vegetative, from a common ancestor.

C-mitosis: Formation of two daughter cells in one cell induced by colchicine resulting in doubling the chromosome number.

Coding strand: DNA has the same sequence as mRNA.

Codominance: Genes when both alleles of a pair are fully expressed in the heterozygote.

Codon: A triplet of nucleotides which represents an amino acid.

Coenocyte: A multinucleate cell resulting from the repeated division of nuclei without cytokinesis.

Coiling: When chromosome cores first become visible in late prophase of mitosis.

Colchicine: An alkaloid chemical induces polyploidy by inhibiting the formation of spindles, delaying the division of kinetochore.

Colchiploidy: Polyploidy induced by colchicine.

Compensating diploids: One normal chromosome that is replaced by two telocentric chromosomes representing either arm or four telocentric chromosomes that compensate for a pair of normal homologues.

Compensating trisomics: A missing normal chromosome is replaced either by two tertiary chromosomes or by a secondary and a tertiary chromosome.

Complement: The group of chromosomes derived from a particular nucleus in gamete or zygote, composed of one, two, or more chromosome sets.

Complementary genes: Two nonallelic gene pairs complement each other in their effect on the same trait. For example, purple-flowered F_1 plants are produced by crossing two white-flowered parents in sweet pea. Selfing of F_1 produces nine purple and seven white; purple arose as the complementary effect of dominant alleles at two different gene pairs.

Complete flower: A flower has sepals, petals, stamens, and pistils.

Complex heterozygous: Special type of genetic system based on the heterozygosity for multiple reciprocal translocations.

Complex locus: A cluster of two or more genes closely linked and functionally related.

Configuration: An association of two or more chromosomes at meiosis and segregating independently of other associations at anaphase.

Congression: The movement of chromosomes onto the metaphase plate led by the orientation of their centromeres at mitosis or meiosis.

Conjugation: The pairing of chromosomes, gametes, or the fusion of pairs of nuclei.

Constitutive heterochromatin: This is differentiated by darker staining during interphase and prophase. In most mammals, constitutive heterochromatin is located in the proximity of the kinetochore. It is an inert state of permanently nonexpressed sequences, usually highly repetitive DNA.

Constitutive mutation: A mutated gene that is usually regulated being always expressed, without regulation.

Continuous spindle fibers: These connect the two polar regions with each other and remain from early prometaphase to early prophase.

Constriction: An unspiralized segment of fixed position in the metaphase chromosomes.

Controlling elements: Originally found and identified in maize by their genetic property, they are transposable units. They may be autonomous (transpose independently) or nonautonomous (transpose only in the presence of an autonomous element).

Convergent coorientation: In a trivalent two chromosomes face one pole while a third one orients to the other pole.

Coorientation: Orientation of kinetochores at metaphase-I.

Coupling: When linked recessive alleles (ab) are present in one homologous chromosome and the other homologous chromosome carries the dominant allele (AB). In the case of repulsion, each homologue contains a dominant and a recessive gene (Ab and aB).

Crossability: The ability of two individuals, species, or populations to hybridize.

Crossing-over: The exchange of chromatid segments of homologous chromosomes by breakage and reunion following synapsis during prophase of meiosis.

Crossing-over map: A genetic map constructed by utilizing crossing-over frequencies as a measure of the relative distance between genes in one linkage group.

Crossing-over value: The frequency of crossing-over between two genes.

Cryptic structural hybridity: Hybrids with chromosomes structurally different but not visible by metaphase-I chromosome pairing.

ct DNA: Chloroplast DNA.

C value: The DNA quantity per genome.

Cytochimera: Tissues contain different chromosome numbers in a single plant organ.

Cytodifferentiation: The sum of processes by which, during the development of the individual, the zygote, specialized cells, tissue, and organs are formed.

Cytogamy: The fusion of cells.

Cytogenetic map: A map showing the locations of genes on a chromosome.

Cytogenetics: A hybrid science that deals with the study of cytology (study of cells) and genetics (study of inheritance). It includes the study of the number, structure, function, and movement of chromosomes and numerous aberrations of these properties as they relate to recombination, transmission, and expression (phenotype) of the genes.

Cytogony: Reproduction by single cell.

Cytokinesis: The division of cytoplasm and organelles in equal proportion. In plants it occurs by the formation of a cell plate while in animals it begins by furrowing. In lower organisms, cytokinesis does not follow after telophase, resulting in multinucleate cells.

Cytokinin: Plant hormones that stimulate shoots in tissue culture.

Cytological hybridization: See *in situ* hybridization.

Cytological maps: The genes located on the maps based on cytological findings such as deficiencies, duplications, inversions, and translocations.

Cytology: The study of cells.

Cytolysis: Breaking up of the cell wall.

Cytomixis: The extrusion of chromatin from one cell into the cytoplasm of an adjoining cell.

Cytoplasm: A portion of the cell apart from the nucleus.

Cytoplasmic inheritance: Inheritance of traits whose determinants are located in the cytoplasmic organelles, also known as non-Mendelian inheritance.

Cytosine: A nitrogenous base, one member of the base pair G-C (guanine and cytosine).

Cytoskeleton: Networks of fibers in the cytoplasm of the eukaryotic cells.

Cytosol: The volume of cytoplasm in which organelles are located.

Cytotaxonomy: The study of natural relationships of organisms by a combination of cytology and taxonomy.

Cytotype: Chromosome numbers other than the standard chromosome complement of that species. For example, *Glycine tomentella* contains accessions with four chromosome numbers ($2n$ = 38, 40, 78, 80).

Daughter cell: The cells resulting from the division of a single cell.

Daughter chromosome: One of the chromatids of mitotic anaphase-I or meiotic anaphase-II.

Decondensation stage: A stage between interphase and prophase of mitosis in which heterochromatin is decondensed for a short time.

Deficiency: The loss of a segment from a normal chromosome that may be terminal or intercalary. The term deletion refers to an internal deficiency.

Deletion: The loss of a chromosomal segment from a chromosome set.

Denaturation of DNA or RNA: Conversion from the double-stranded to the single-stranded state accomplished by heating.

Deoxyribose: The four-carbon sugar that characterizes DNA.

Desynapsis: The premature separation of homologous chromosomes during diakinesis/metaphase-I; often genetically controlled.

Detassel: Removal of male inflorescence (tassel) from maize.

Diakinesis (meiosis): Chromosomes continue to shorten and thicken. This is a favorable stage to count chromosomes. Chromosomes lie well-spaced out uniformly in the nucleus and often form a row near the nuclear membrane.

Dicentric chromosome: A chromosome with two kinetochores.

Dichogamous: Protandry; male and female organs mature at different times allowing cross-pollination.

Didiploid: Two different diploid chromosome sets present in one cell or organism (allotetraploid).

Differentiation: It occurs in cell, tissue, and organ cultures where a plant develops from meristematic cells.

Diffuse kinetochore: Centromere activity is spread over the surface of the entire chromosome in certain insects of the order *Hemiptera* and in plant *Luzula*.

Diffuse stage: Prophase stage (meiosis) in which the chromosome become reorganized; they even disappear.

Dihybrid cross: A genetic cross involving two genes.

Diisosomic: An individual which has a pair of homologous isochromosomes from one arm of a particular chromosome.

Diisotrisomic: An individual that lacks one chromosome but carries two homologous isochromosomes of one arm of a particular chromosome.

Dikaryon: A dinucleate cell.

Dioecious: A plant bears either staminate or pistillate flowers.

Diplochromosome: Chromosome which has divided twice during the resting stage without division of its centromere.

Diplohaplontic: This phenomenon occurs in higher plants and in many algae and fungi. Life cycle of a typical plant contains spore-bearing ($2n$) and gamete-bearing (n) stages.

Diploid: An organism with two sets of chromosomes (two complete genomes). For example, barley is a diploid with $2n = 2x = 14$.

Diploidization: In polyploidy, chromosomes exhibit diploid-like meiosis; genetically controlled.

Diploidy: The presence of two homologous sets of chromosomes in somatic cells.

Diploid-like meiosis: An allopolyploid showing bivalent chromosome pairing that, sometimes, is under genetic control.

Diplonema (meiosis): A stage at which sister chromatids begin to repulse at one or more points in a bivalent.

Diplontic: Mature individuals are diploid by mitotic divisions and differentiations. Meiosis produces only haploid gametes. Diplophase is more prominent than the haplophase in humans, higher animals, and some algae.

Diplontic selection: Diploid cells overgrow the aneuploid cells in culture.

Diplospory: The embryo sac is formed from a megasporocyte as a result of abnormal meiosis.

Direct tandem inversions: This is due to two successive inversions involving chromosome segments directly adjacent to each other.

Disjunction: Migration (separation) of chromosomes to opposite poles at meiotic anaphase.

Disomic inheritance: This occurs by the association of chromosomes in bivalents at meiosis.

Dispermic: An egg fertilized by two male gametes.

Distal chromosome: A part of a chromosome, which is relative to another part, is farther from the kinetochore (opposed to proximal).

Distant hybridization: The crossing of species of different genera.

Ditelomonotelosomic: An individual that has a pair of telocentric chromosomes for one arm and a single telocentric chromosome for the other arm.

Ditelosomic: An individual that has two telocentrics of one chromosome arm.

Ditelotrisomic: An individual that has two telocentric chromosomes from one arm plus one complete chromosome of the homologue.

Ditertiary compensating trisomics: An individual with a compensating trisomics chromosome in which a missing chromosome is compensated by two tertiary chromosomes.

Dominance: When one member of a pair of alleles manifests itself more or less, when heterozygous, than its alternative allele.

DMSO: Dimethyl sulfoxide.

DNA: Deoxyribose nucleic acid. All hereditary information is encoded.

DNase: An enzyme that attacks bonds in DNA.

Double crossing-over: The production of a chromatid in which crossing-over has occurred twice.

Double cross: Crossing two single-cross F_1 hybrids.

Double fertilization: The union of one sperm nucleus with the egg nucleus and one with polar nuclei (secondary nucleus) to form an embryo and endosperm in flowering plants.

Double heterozygote: Heterozygote in respect of two genes.

Double haploid: A diploid plant obtained from spontaneous or induced chromosome doubling of a haploid cell or plant, usually after anther or microspore culture.

Double reduction: Chromatid segregation.

Drosera scheme pairing: A triploid hybrid showing $2x$ bivalent + $1x$ univalent configurations at diakinesis or metaphase-I.

Duplex genotype: AAaa.

Duplication: When an extra piece of chromosome segment is attached to either the same chromosome (homologous) or transposed to another member (nonhomologous) of the genome.

Dyad: Two cells product of first meiosis.

Dysploid: A species with differing basic chromosome numbers (e.g., x = 5; 6; 7; 8; etc.). Basic chromosome numbers in dysploid are stable.

Dysploidy: Abnormal polyploidy.

Ectopic pairing: Chromosome pairing between an allele and nonallelic (ectopic) homologous sequence due to some homology searching mechanism.

Egg: The female gamete as opposite to sperm.

Egg apparatus: This includes the egg cell and two synergid cells at the micropyle end of the embryo sac.

Electron microscope: A microscope that uses an electron beam as a light source.

Emasculation: Removal of anthers from a flower.

Embryo culture: The excision of young embryos under aseptic conditions and transferring them on suitable artificial nutrient media for germination or callusing.

Embryogenic cells: Cells that have completed the transition from somatic state to one in which no further exogenously applied stimuli are necessary to produce the somatic embryos.

Embryology: To study the development of reproductive organs in an organism.

Embryo: Rudimentary plant in a seed.

Embryoid: An embryo-like structure observed in cell and tissue cultures.

Embryo sac: This is a mega gametophyte of flowering plants; includes egg, synergids, polar nuclei, and antipodals.

Endogamy: Sexual reproduction in which the mating partners are more or less closely related.

Endomitosis: Separation of chromosomes but two sister chromatids lie side by side and retention within a nucleus.

Endoplasmic reticulum: The outer membrane of the nuclear envelope.

Endopolyploidy: Polyploidy that arises from repeated chromosome replication without cytokinesis.

Endoreduplication: Chromosomes replicate time after time without condensation and mitosis, and appear as bundles of multiple chromatids, for example, polytene chromosomes.

Endosome: A nucleolus-like organelle that does not disappear during mitosis.

Endosperm: Triploid tissue that originates from the fusion of a sperm nucleus with the secondary nucleus and nourishes the embryo during early growth of the seedling.

Endothelium: The inner epidermis of the integument, next to the nucellus, where cells become densely cytoplasmatic and secretory. It is also known as integumentary tapetum.

Endothesium: A layer of cells situated in the pollen-sac wall situated below the epidermis.

Enneaploid: A polyploid with nine chromosome sets ($9x$).

Epistasis: Dominance of one gene over a nonallelic gene.

Equational division: A division of each chromosome into exact longitudinal halves which are incorporated equally into two-daughter nuclei, also known as mitosis.

Equational plate: An arrangement of the chromosomes in which they lie approximately in one plane, at the equator of the spindle; seen during mitotic and meiotic metaphase.

EST: Expressed sequence tag.

Euchromatin: The decondensed region of a chromosome containing the majority of the genes; stains lightly.

Euchromatization: The spontaneous or induced change of heterochromatin into euchromatin.

Euchromosome: A chromosome showing the typical features of the standard complement of a given species.

Euhaploid: A haploid genome showing no deviating number of chromosomes compared with the standard genome of the species.

Eukaryote: An organism with a true nucleus.

Eumeiosis: Normal occurring meiosis.

Euploid: An organism possessing chromosomes in an exact multiple of the basic (x) chromosome number such as $2x$, $3x$, $4x$, and so on.

Eusom: A plant showing each member of the chromosome complement with the same copy number.

Euspory: The typical sporogenesis with normal flow of reduction division-reduced apomixis.

Eutelomere: A weakly staining subterminal segment adjacent to the protelomere.

Expressivity: The degree of phenotypic effect of a gene in terms of deviation from the normal phenotype.

Extrachromosomal inheritance: Inheritance that is not controlled by nuclear genes but by cytoplasmic organelles.

Extra chromosome: B-chromosome.

Extranuclear genes: Genes located in the cytoplasmic organelles such as mitochondria and chloroplasts.

F_1, F_2, etc.: Filial generations. The offspring resulting from experimental crossing (F_1) and selfing (F_2).

Facultative apomixis: The plant produces seed both apomictically and sexually.

Facultative heterochromatic: Chromatin which is heterochromatinized, may or may not be condensed in interphase.

Female: Female person, animal, or plant. In plants, having pistil and no stamens; capable of being fertilized and bearing fruits.

Fertilization: Fusion of a male sperm with an egg cell (formation of zygote).

Feulgen staining: A cytological colorless stain, utilizes Feulgen or Schiff's reagent. Chromosomes are colored, cytoplasm, and nucleolus are colorless and are specific for DNA.

Filament: The stalk of a stamen.

Filiform apparatus: A complex wall outgrowth in a synergid cell.

FISH: Fluorescence *in situ* hybridization.

Fission: The division of one cell by cleavage into two daughter cells, or a chromosome into two arms.

Fixation: The killing of cells or tissues by chemicals in such a way that the internal structure of a cell is preserved.

Fluorescence *in situ* hybridization (FISH): A cytological technique for the detection of repetitive DNA sequences such as ribosomal DNA sequences (NOR sequences).

Fluorescence microscopy: The principle of fluorescence microscopy is based on the light transmitted or reflected by the specimen. The specimen is stained with a fluorochrome (a dye that emits a longer wavelength when exposed to the shorter wavelength of light). The fluorescent region of a specimen becomes bright against a dark background.

G_1: The period of the eukaryotic cell cycle between the last mitosis and the start of DNA replication (synthesis).

G_2: The period of the eukaryotic cell cycle between the end of DNA replication (synthesis) and the start of the next mitosis.

G-Bands: Bands observed in the chromosomes after Giemsa staining. The chromosomes are pretreated with a dilute trypsin solution, urea, or protease. Bands have been used to identify human and plant chromosomes.

Gamete: A sex cell that contains half the chromosome complement of the zygote formed after meiosis that participates in fertilization.

Gametic chromosome number: *n*.

Gametocyte: A cell that produces gametes through division; a spermatocyte or oocyte-germ cell.

Gametogenesis: The formation of male and female gametes.

Gametophyte: A part of the life cycle of a plant which produces the gametes where there is an alternation of generations.

G-banding: A cytological technique which exhibits a striated pattern in human chromosomes.

Gene: The unit of inheritance located mainly in chromosomes at fixed loci and contains genetic information that determines characteristics of an organism.

Gene-for-gene theory: Plant–pathogen interaction.

Gene location: Determination of physical or relative distances of a gene on a particular chromosome.

Gene pool: The reservoir of different genes of a certain plant species (lower and higher taxa) available for crossing and selection; divided based on hybridization into primary, secondary, and tertiary gene pools.

Generative nucleus: A product (two-daughter nuclei) of division of primary nucleus in the pollen grain. One usually divides into two sperm nuclei in the pollen tube while a vegetative nucleus does not divide again.

Genetic: A property obtained by an organism by virtue of its heredity.

Genetic code: The correspondence between triplets in DNA (or RNA) and amino acids in proteins.

Genetic combination: For n pairs of homologous chromosomes, only $1/2n-1$ gametes will contain parental combination.

Genetic marker: An allele used to identify a gene, or a chromosome, or a chromosome segment.

Genetic map: The linear arrangement of gene loci on a chromosome determined from genetic recombination.

Genetic mapping: The assignment of genes to specific linkage groups and determination of the relative distance of these genes to other known genes in that particular linkage group.

Genetics: The science which deals with heredity and variation.

Genome: A complete haploid set of chromosomes of a species.

Genome doubling: Chromosome doubling of F_1 hybrid for producing auto- or allo polyploidy.

Genome segregation: The segregation of sets of chromosomes (genomes) during mitosis in the somatic tissues that occurs in polyploid species and also in some wide crosses.

Genome symbol: Assignment of symbol to species, usually by capital letter, by cytogenetic, and molecular methods.

Genomic *in situ* hybridization (GISH): A cytological technique by which a genome-specific probe is used. DNA sequence specific to a particular genome serves as a marker to identify genome, translocations, and alien addition and substitution lines.

Genotype: The description of the genes contained by a particular organism for the character or characters—the entire genetic constitution.

Genotypic control: The control of chromosome behavior by the hereditary properties of the organism.

Genotypic ratio: Proportion of different genotypes in a population.

Germ cell: Gamete.

Golgi apparatus: All golgi bodies in a given cell are not like staining bodies usually engaged in secretion.

Golgi bodies: A cell organelle made up of closely packed broad cisternae and small vesicles.

Guanine: A nitrogenous base, one member of the base pair G-C (guanine and cytosine).

Gynoecium: The female portion of the flower—stigma, style, and ovary.

Haploid: An organism containing a gametic (n) chromosome number.

Haplotype: The particular combination of alleles in a defined region of some chromosome, in effect the genotype in miniature.

Hemialloploid: Segmental allopolyploid.

Hemigamy: Sometimes the sperm nucleus does not unite with the egg nucleus and divides independently and simultaneously. The embryo sac nuclei are unreduced. Chimeric embryos (somatic + half the somatic chromosome) develop.

Hemiploidy: An individual with a haploid chromosome number.

Hemizygous: A gene allele opposite to a deficiency.

Hermaphroditic: Flowers include male and female sexes.

Hereditary: Transmissible from parent to offspring.

Hereditary determinant: Gene.

Heredity: The process which brings about the biological similarity and dissimilarity between parents and progeny.

Heterochromatin: The region of a chromosome that takes differential stain. These regions contain condensed chromatin.

Heterochromosome: Any chromosome that differs from the autosomes in size, shape, and behavior.

Heterogenetic: Chromosome pairing between more or less different genomes in amphidiploids.

Heterogenetic chromosome pairing: The F_1 hybrids of genomically divergent species lack chromosome affinity and form only a small number of loosely synapsed bivalents.

Heterokaryotype: A chromosome complement that is heterozygous for any sort of chromosome mutations.

Heteromorphic bivalent: A bivalent in which one of the chromosomes differs structurally. The chromosomes may differ in size or kinetochore positions.

Heteroploid: An individual containing a chromosome number other than the true monoploid or diploid number.

Heteropycnosis: Differential spiralized regions of chromosomes. Positive at interphase or early prophase, negative at late prophase or metaphase.

Heterozygous: A zygote derived from the union of gametes unlike in respect of the constitution of their chromosomes or from genes in a *heterozygote* (e.g., *Aa*). Also, used in the case of a hybrid for interchanges, duplications, deficiencies, and inversions—the opposite of *homozygotes*.

Hexaploid: A polyploid with six sets of basic chromosomes.

Highly repetitive DNA: This is equated with the satellite DNA which is composed of many tandem repeats (identical or related) of a short basic repeating unit. (= simple sequence DNA).

Hilum: The scar on the surface of seed revealed by abscission of the funicle; the point of attachment of the seed to the pod.

Histology: Study of tissues.

Holocentric: Diffused kinetochore. Every point along the length of chromosome exhibits centromeric activity.

Holandric genes: The genes located on Y chromosomes.

Homoeobox: A short stretch of nucleotides whose base sequence is virtually identical in all the genes that contain it. It determines when particular groups of genes are expressed during development.

Homoeologous chromosomes: Partially homologous chromosomes; for example, in wheat ABD genome chromosomes are known as homoeologous.

Homogametic sex: An organism has $2x$ chromosomes and autosomes.

Homogenetic chromosome pairing: Synapsis of homologous chromosomes of the same genome.

Homologous chromosomes: Like chromosomes showing synapsis all along their length during pachynema. For n pairs of homologous chromosomes, only $1/2n-1$ gametes will contain parental combination.

Homozygote: An organism having like alleles at corresponding loci on homologous chromosomes (AA, aa).

Hotspot: A site in the chromosome at which the frequency of mutation (or recombination) is very much increased.

Hyperploid: A diploid organism with an extra chromosome.

Hypertriploid: An individual containing chromosome numbers higher than $3x$.

Hypoploid: A diploid organism lacking an extra chromosome.

Ideotype: An ideal plant form formulated to assist in reaching selection goals.

Idiogram: A diagrammatic representation of chromosome morphology of an organism.

Included inversions: A segment that is part of an inverted segment is inverted once again.

Independent assortment: The random distribution to the meiotic products (gametes) of genes located on different chromosomes. A dihybrid individual with $RrYy$ genotype will produce four gametic genotypes (RY, Ry, rY, and ry) in equal proportions.

Independent inversions: Such inversions occur in independent parts of the chromosomes and the two resultant inverted segments are separated from each other by an uninverted (normal) chromosome segment.

Inheritance: The transmission of genetic information from parents to progenies.

In situ **hybridization:** A cytological technique conducted by denaturing the DNA of cells squashed on a microscope slide so that a reaction is feasible with an added single-stranded DNA or RNA which is labeled, and following hybridization the gene is located on a certain site of a chromosome.

Intercalary: Chromosomal segments located beside the terminal region of chromosomes.

Interchange (reciprocal translocation): The reciprocal exchange of terminal segments of nonhomologous chromosomes.

Interchange trisomics: An additional chromosome to a diploid set, which is composed of two different chromosomes (translocation).

Interkinesis: The time gap between cell division I and II. It is short or may not occur at all.

Interphase: The period between mitotic cell division (or intermitotic period) which is divided into G_1, S, and G_2.

Interstitial segment: A chromosome region between the centromere and a site of rearrangement.

Intrachromosomal: Within a chromosome.

Intragenomic pairing: Pairing of chromosomes in monohaploid.

Inversion: Reversal of gene order within a chromosome, may be paracentric which does not include kinetochore or pericentric which involves kinetochore in the inverted region.

Inversion heterozygote: One chromosome carries a normal gene sequence while the other chromosome contains an inverted gene sequence; may be paracentric or pericentric.

Isochromocentric: Nuclei showing as many chromocenters as chromosomes.

Isochromosome: A chromosome with two identical arms; it is usually derived from telocentric chromosomes.

Isodicentric chromosome: An abnormal chromosome containing a duplication of part of the chromosome including the centromere; the resulting chromosome contains two centromeres and a point of symmetry that depends on the position of the breakpoint.

Isosomic: Individuals showing isochromosome.

Isotelocompensating trisomic: A compensating trisomic; a missing chromosome is compensated for one telocentric and one tertiary chromosome.

Isotertiary compensating trisomic: A compensating trisomic; a missing chromosome is compensated for by one isochromosome and one tertiary chromosome.

Isotrisomic: An individual containing one isochromosome above the normal chromosome complement.

Karyoevolution: Evolutionary change in the chromosome set, expressed as number and structure of chromosomes.

Karyogamy: The fusion of two gametic nuclei.

Karyogram: Ideogram.

Karyokinesis: The process of nuclear division.

Karyology: Study of the nucleus.

Karyostatasis: The stage of cell cycle in which there is no visible dividing activity of the nucleus, but metabolic and synthetic activity.

Karyotype: The usual chromosome number, size, and morphology of the chromosome set of a cell of an organism. The complete set of chromosomes. For example, barley contains $2n = 2x = 14$.

kb: An abbreviation for 1,000 base pairs of DNA or 1,000 bases of RNA.

Kinetochore: The region of a chromosome attached to a spindle fiber during movement of the chromosomes at mitosis and meiosis.

Kinetochore-misdivision: Division of kinetochore crosswise instead of lengthwise particularly during meiosis.

Kinetochore-orientation: The process of orientation of kinetochore during the prometaphase of mitosis and meiosis.

Knob: A heterochromatic darkly stained region that may be terminal or intercalary. It is often used as a landmark to identify specific chromosomes.

Lagging: Delayed migration of chromosome from the equatorial plate to the poles at mitotic or meiotic anaphase and consequently excluded from the daughter nuclei producing aneuploid gametes.

Lampbrush chromosome: A special type of chromosome found in the primary oocyte nuclei of vertebrates and invertebrates. Such chromosomes are due to prolonged diplonema. Chromosomes exhibit a fuzzy appearance due to paired loops. These loops vary in size and shape from chromosome to chromosome and have been used to construct cytological maps.

Leptonema: Chromonemata become distinct from one another and appear as very long and slender threads. Nuclei contain a large nucleolus and a distinct nuclear membrane.

Lethal gene: Any gene in which a lethal mutation can be obtained (deletion of the gene).

Linkage: Association of qualitative and quantitative traits in inheritance due to location of marker genes in proximity on the same chromosome.

Linkage drag: The inheritance of undesirable genes along with a beneficial gene due to their close linkage.

Linkage group: This includes all loci that can be connected (directly or indirectly) by linkage relationships; equivalent to haploid chromosome. For example, barley has seven linkage groups corresponding to seven haploid chromosome numbers. All genes located in one chromosome form one linkage group.

Linkage map: A graphical representation of the positions of genes along a chromosome.

Linkage value: Recombination fraction: the proportion of crossovers vs. parental types in a progeny. It may vary from 0%–%.

Localized kinetochore: A chromosome carrying a normal permanent kinetochore localized to which the spindle fiber is attached during chromosome separation.

Locus: The position of a gene in a chromosome.

Maceration: The artificial separation of the individual cell from a tissue by enzyme treatment.

Map distance: This is measured in cM (centi Morgan) = percent recombination.

Marker: An identifiable physical location on a chromosome, determined by inheritance.

Maternal inheritance: Inheritance of a trait through a female.

Median centromere: A centromere that is located midway on chromosomes resulting in two equal arm lengths.

Megabase (Mb): Unit length for DNA fragments equal to 1 million nucleotides and roughly equal to 1cM.

Megagametophyte: Female gametophyte; embryo sac within the ovule of angiosperms.

Megaspore: A gametophyte bearing only female gametes.

Megaspore mother cells (MMCs): Produce eight nuclei (megaspores) after one of four haploid macrospores becomes functional in the embryo sac. Two mitotic divisions produce eight nuclei (one egg, two synergids, three antipodals, and two polar nuclei).

Meiosis: A phenomenon which occurs in the reproductive organs of male and female organisms, producing gametes with the haploid (n) chromosome constitution. The first division is reductional and the second division is equational.

Meiotic duration: The time required to complete the cell cycle from prophase to telophase under certain conditions.

Meiotic segregation: The random orientation of bivalents at the equatorial plate at metaphase-I regulates the segregation of paternal and maternal chromosomes to the daughter nuclei.

Meristem: An actively growing tissue that produces cells which undergo differentiation to form mature tissues.

Mesopolyploid: Individuals have undergone a moderate aneuploid-induced base number shift. Apomicts may contain complete or nearly complete duplicate sets of genes.

Metacentric chromosome: A chromosome with a median kinetochore (both arms are equal in length).

Metakinesis: Prometaphase.

Metaphase (mitosis): Chromosomes are arranged at the equatorial plate. Complete breakdown of the nuclear membrane and disappearance of the nucleolus occur.

Metaphase arrest: The arrest of cell division at mitotic metaphase, usually by application of specific agents.

Metaphase-I: Bivalents are arranged at the equatorial plate with their kinetochores facing its poles. Chromosomes reach to their maximum contractions, nuclear membrane and nucleolus disappear, and spindle fibers appear.

Metaphase plate: The movement and arrangement of all the chromosomes midway between the two poles of the spindle is called the metaphase plate.

Metastatis: The ability of tumor cells to leave their site of origin and migrate to other locations in the body where a new colony is established.

Microchromosomes: Tiny dot-like chromosomes that are too small for their centromere and individual arms to be resolved under the light microscope.

Microgametogenesis: The development of pollen grains.

Microgametophyte: The male gametophyte of a heterosporous plant.

Micronucleus: A nucleus separate from the main nucleus formed at telophase by one or more lagging chromosomes or fragments.

Micropyle: An opening of the ovule.

Microsomes: The fragmented pieces of endoplasmic reticulum associated with ribosomes.

Microspore: A gametophyte producing only male gametes—a pollen grain.

Microsporocyte: A cell that differentiates into a microspore.

Microspore Mother Cells: Diploid cells in the stamens (pollen mother cells, PMCs) give rise to haploid microspores after meiosis.

Microtubules: Manipulation of chromosomes in eucaryotes is due to microtubules. They are organized into spindle-shaped bodies, disappear at the end of cell division.

Minichromosomes: Small chromosomes with kinetochores, totally heterochromatic.

Minute chromosomes: Usually very tiny chromosome segments as a result of chromosome aberrations.

Misdivision: Aberrant chromosome division in which transversal but no longitudinal separation of the centromere occurs.

Mitochondrion: A cell organelle found in the cytoplasm. Mitochondrial DNA is responsible for extra nuclear inheritance. It is the site of most cellular respiration.

Mitosis: A process of cell division known as equational division because the exact longitudinal division of the chromosomes into two chromatids occurs. The precise distribution into two identical daughter nuclei leads to the generation of two cells identical to the original cell from which they were derived.

Mitotic anaphase: Daughter chromosomes are pulled to opposite poles by the spindle fibers.

Mitotic apparatus: An organelle consisting of three components: (1) the asters which form the centrosome; (2) the gelatinous spindle; and (3) the traction fibers, which connect the centromeres of the various chromosomes to either centrosome.

Mitotic crossing-over: Somatic crossing-over.

Mitotic cycle: The sequence of steps by which the genetic material is equally divided before the division of a cell into two daughter cells opens.

Mitotic index (MI): The fraction of cells undergoing mitosis in a given sample; usually the fraction of a total of 1,000 cells that are undergoing division at one time.

Mitotic inhibition: Induced or spontaneous inhibition of mitotic division.

Mitotic prophase: Chromosomes begin to reappear, are uniformly distributed in the nucleus, and are more or less spirally coiled and seem longitudinally double.

Mitotic metaphase: Chromosomes are shrunk to the maximum limit, kinetochores move onto the equatorial plate.

Mitotic poison: Any substance that hampers proper mitosis.

Mitotic recombination: A type of genetic recombination that may occur in somatic cells during their preparation for mitosis in both sexual and asexual organisms.

Mitotic telophase: By the end of anaphase spindle fibers disappear and chromosomes at each pole form a dense ball. Nucleolus, nuclear membranem, and chromocenters re-emerge. Chromosomes lose their stainability.

Mixoploidy: A failure in chromosome migration results in the formation of multiploid microsporocytes.

Monocentric: One centromere per chromosome.

Monogenic character: A character determined by a single gene.

Monohybrid cross: A genetic cross which involves only one pair of genes.

Monoploid: An organism having basic (x) chromosome number.

Monosomics: Where one of the chromosomes is missing from the normal chromosome complement ($2n-1$).

Monotelotrisomics: An individual with 2n+1 telocentric chromosome.

Monozygous twins: This is also known as identical twins. Twins arise from a single zygote formed by the fertilization of a single egg by a single sperm. In this case the developing egg cleaves at an early division forming two separate embryos instead of the usual one.

Morphogenesis: The differentiation of structures during development.

Mother-cell: A diploid nucleus which by meiosis yields four haploid nuclei—the microspore or pollen mother-cell (PMC) and the megaspore or embryo sac mother-cell in flowering plants.

mt DNA: Mitochondrial DNA.

Multiform aneuploidy: An organism contains cells with different aneuploid chromosome numbers producing the tissues with chromosome mosaicism.

Multiple factors: Many genes involved in the expression of any one quantitative character.

Multivalent: An association of more than two completely or partially homologous chromosomes during meiosis, generally observed in chromosomal interchanges and in autopolyploidy.

Mutant: An organism carrying a gene resulted from mutation.

Mutation: Any heritable change in the genetic material (sequence of genomic DNA) not caused by segregation or genetic recombination, may occur spontaneously or induced experimentally.

A detectable and heritable change in the genetic material, transmitted to succeeding generations.

n: haploid chromosome number symbol.

N-bands: This procedure was originally developed to stain nucleolus organizer chromosomes. Chromosomes are treated at low pH (4.2 ± 0.2) in 1 M NaH_2Po_4 for 3 min at high temperature (94 ± 1˚C). Chromosomes are stained with Giemsa after incubation and washing.

Neocentric: Where the activity of the centromeres is exceptionally transferred to the ends at an abnormal meiosis.

Neocentromeres: Centromere region is replaced by a secondary center of movement; chromosome ends move first during anaphase-I of meiosis.

Neopolyploid: Polyploid individuals of recent origin.

Neurospora: The common *Neurospora crassa* is a bread mold. The normal vegetative phase is a branching haploid individual with hyphae which form a spongy pad, the mycelium. Mycelium fuse with each other but nuclei do not (heterokaryon), although mixed and remain in the same cytoplasm. Asexual reproduction occurs by repeated mitosis of nuclei or by the mitotic formation of special haploid spores (conidia). Conidia can form a new mycelium.

Nondisjunction: The failure of separation of paired chromosomes at mitotic and meiotic anaphase resulting in one daughter cell receiving both and the other daughter cell receiving none of the chromosomes.

Nonchromosomal: This shows a non-Mendelian inheritance; the genes are extrachromosomally located in chloroplasts and mitochondria.

Nonhomologous chromosome association: Pairing of nonhomologous chromosomes at meiotic metaphase-I; usually observed in haploid.

Noncoorientation: In a quadrivalent, two nonhomologous centromeres are on opposite sides and centromeres of two translocated chromosomes are stretched between the other two and not attached to the opposite poles. Noncooriented quadrivalent configuration results in a 3:1 chromosome segregation, it always produces unbalanced gametes and generates trisomics in the progenies.

Nondisjunction: This can occur during mitosis and meiosis. Chromosomes fail to separate at metaphase resulting in their passage to the same pole. This may occur due to failure of chromosome synapsis, or from multivalent formation.

Nonhomologous: Chromosomes carrying dissimilar chromosome segments and genes that do not pair during meiosis.

Nonlocalized centromeres: The spindle attachment is not confined to a strictly localized chromosome area.

Northern blotting: A molecular biology technique for transferring RNA from an agarose gel to a nitrocellulose filter on which it can be hybridized to a complementary DNA.

Nucellus: Ovule tissues internal to the integuments and surrounding the embryo sac.

Nuclear division: The division of the cell nucleus by mitosis, meiosis, and amitosis.

Nuclear envelope: A doubled-layer outer membrane of the nucleus which separates nucleus and the cytoplasm of the cell in eukaryotes.

Nuclear fragmentation: The degeneration of the nucleus by partition the nucleus into more or fewer different parts.

Nuclear pore: Holes in the nuclear envelope; assumed to be used for transport of macromolecules.

Nucleic acid: A large molecule composed of nucleotide subunits.

Nucleolar chromosome: The chromosome that carries the nucleolus organizer region.

Nucleolar dominance: Amphiplasty.

Nucleotype: The gross physical characterization of the nucleus; its mass and particularly the amount of DNA content.

Nucleolus: A spherical body found inside the nucleus is associated with a specific chromosomal segment (the nucleolus organizer); it is active from telophase to the following prophases in the site of ribosomal RNA (rRNA) synthesis.

Nucleolus organizer region: The chromosome region which is associated with the nucleolus, responsible for the formation of the nucleolus and contains ribosomes; referred to as the NOR.

Nucleosome: The basic structural subunit of chromatin, consisting of ~200 bp of DNA and an octamer of histone of proteins.

Nucleus: A spheroidal, most important body within the cells of eukaryotes which contains chromosomes.

Nulliplex: In polyploidy, one recessive allele is carried by all chromosomes (*aaaa*).

Nullisomics: An individual lacking one chromosome pair from the normal chromosome complement ($2n-2$).

Nullitetrasomics: In allohexaploid wheat, one pair of homologous chromosomes is missing and is substituted by two pairs of homologous chromosomes.

Null mutation: Physical deletion of a gene.

Oöcyte: The cell which becomes the egg by meiosis.

Oögonium: A mitotically active cell which gives rise to oöcytes.

Organelle: Any structure of characteristic morphology and function in a cell, such as nucleus, mitochondrion, and chloroplast.

Overlapping inversions: A part of an inverted segment being inverted a second time together with an adjacent segment that was not included in the first inversion segment.

Ovum: An unfertilized egg—ovule, egg.

Pachynema: Synapsis of homologous chromosomes is completed. Chromosomes are noticeably thicker and shorter than those in leptonema. Nucleolus is clearly visible and certain chromosomes may be attached to it.

Pairing: Chromosome pairing—synapsis.

Pairing segment: The segment of chromosomes (X and Y chromosomes) which pairs and crosses over (pairing segment). The remaining segments which do not pair, are known as differential segments.

Paleopolyploid: Ancient polyploids; all their $2x$ progenitors apparently have become extinct.

Paracentric inversion: A type of intrachromosomal structural change which does not include the kinetochore of the chromosome.

Paranemic coiling: Chromatids are easily separated laterally.

Pentaploid: $5x$.

Pericentric inversion: A type of inversion that includes the kinetochore within the inverted segments.

Phenotype: The observable external appearance of an organism produced by the interaction of genotype and the environment.

Plastogene: Genes located in the plastids showing non-Mendelian inheritance.

Plectonemic coiling: Chromatids cannot be separated without unwinding the coil.

Ploidy: The number of chromosome sets per cell of an organism as haploid, diploid, and polyploid.

Point mutation: A change in single base pairs (gene mutation).

Polycentromeres: Each chromosome is attached by many spindle fibers.

Polyhaploid: Haploid (n) individuals arising from polyploid species.

Polymitosis: Occurence of a series of mistoses in rapid succession with or without division of chromosomes.

Polymorphism: The occurrence of two or more genetically different types in the population—allelic variation. In molecular terms, changes in DNA influence the restriction pattern.

Polyploid: An organism containing more than two sets of chromosome.

Polyploidization: Induction of polyploids artificially or naturally from haploid and diploid plants.

Polyspory: This is characterized by numerous cytological abnormalities because complete sets of asynchronously-expressed genes are not present. It is associated with the frequent occurrence of aneuploid series formation and stabilization (*Poaceae*).

Polytene chromosome: A giant chromosome consisting of many identical chromosomes closely associated along their entire length (endomitosis). In the *Drosophila melanogaster* salivary gland chromosomes, about 5,000 bands can be recognized. Homologous polytene chromosomes pair along the length in the somatic cells (somatic pairing). Any chromosomal structural changes could be determined by differences in the pairing pattern. Changes in appearance of bands in certain tissues occur at different times, either condensing or expanding to form "puffs."

Primary constriction: Synonymous with centromere or kinetochore.

Primary trisomics: An individual with a normal chromosome complement plus an extra complete standard chromosome is designated as a simple primary trisomic.

Primer: A short sequence, often RNA that is paired with one strand of DNA and provides a free 3'-OH end at which a DNA polymerase starts.

Probe: Single-stranded DNA or RNA molecules of specific base sequence, labeled either radioactively or immunologically, used to detect the complementary base sequence by hybridization.

Prochromosomes: Small condensed chromosomes appearing in the resting nuclei of certain plant species; strictly not chromocenters or heterochromatin.

Prophase (mitosis): This is the first stage of mitosis; cells prepare for cell division. Coiling of chromosomes (condensation) occurs and chromosomes appear as a thread-like structure. Prophase includes splitting of chromosomes longitudinally into two duplicates (chromatids) and the centrosome and dissociation of each half to opposite side of the nucleus, the synthesis of mitotic apparatus, the disappearance of nucleolus and the beginning of the breakdown of the nuclear membrane.

Pseudogamy: In plants where an embryo develops from a diploid or haploid egg cell with a sperm penetrated but without fertilization and the egg cell nucleus remains functional.

Pseudogenes: These are inactive genes but are stable components of the genome derived by mutation of an ancestral active gene.

Pseudoisochromosome: Chromosome ends are homologoues due to reciprocal translocation; chromosomes pair at meiotic metaphase-I like an isochromosome but the interstitial segments (proximal to kinetochore) are nonhomologous.

Purine: A nitrogen-containing, single-ring, basic compound that occurs in nucleic acids. The purines in DNA or RNA are adenine and guanine.

Pyrimidine: A nitrogen-containing, double-ring, basic compound that occurs in nucleic acids. The pyrimidines in DNA are cytosine and thymine; in RNA, cytosine, and uracil.

Q: Human chromosome nomenclature for the long arm.

Q-bands: Caspersson et al. (1968) showed that chromosomes of many plant species stained with quinacrin mustard exhibit bright and dark zones under ultraviolet light. This method was later used for the identification of individual human chromosomes (A-T rich region of a chromosome).

Quadripartitioning: During meiosis, cytokinesis does not occur after meiosis I and is postponed until after the second division.

Quadrivalent: An association of four chromosomes generally observed in a reciprocal translocation or in autotetraploids.

Quartet: The four nuclei produced at the end of meiosis (= tetrad).

Quasibivalent: Pseudobivalent—bivalent like association during meiosis due to stickiness rather than by chiasmata.

R-bands: The lightly stained G-bands become darkly stained after R-banding staining. R-bands are obtained after treating chromosomes at low pH (4–4.5) at 88°C incubation in NaH_2Po_4 (1 M). R-bands are reversible if pH is adjusted to 5.5–5.6.

Radioautography: A photographic technique in which a radioisotope is taken in by a cell and can be traced with a sensitive photographic film.

Reassociation of DNA: The pairing of complementary single strands of DNA to form a double helix.

Recombination: The occurrence of new combinations of genes in hybrids, not present in the parents, due to independent assortment and crossing-over in crosses between genetically different parents.

Reduction: A process which halves the somatic chromosome number at meiosis; segregation of homologous includes a member of each set but these chromosomes are no longer the unaltered parental chromosomes.

Regulatory gene: This codes for an RNA or protein product whose function is to control the expression of the other gene.

Relational coiling: Chromatids are twisted about each other during early prophase.

Renaturation: The return of a denatured nucleic acid or protein to its native configuration.

Renner complexes: In *Oenothera*, several genetic factor complexes are found and these complexes segregate as a whole in meiosis, with each gamete carrying one or the other. Certain complexes are lethal in gametes, (gametophytic lethal) while some combinations are lethal zygotes (zygotic lethal). Specific names were assigned to these complexes by Renner and they are known as Renner complexes.

Repetitive DNA: Nucleotide sequences that are present in the genome in many copies.

Reproductive isolation: It is an isolating mechanism of evolution where a population is isolated by several genetically controlled mechanisms which prevent gene exchange between two populations.

Resting stage: Interphase; any nucleus not undergoing division. However, the nucleus is very active metabolically.

Restitution: The spontaneous rejoining of broken ends of chromosomes to produce the original chromosome configuration.

Restitution nucleus: A single nucleus with unreduced chromosomes is produced through failure of one of the divisions of meiosis.

Restriction enzyme: This recognizes specific short sequences of (usually) unmethylated DNA and cleaves the duplex randomly.

Restriction fragment length polymorphism (RFLP): Inherited differences in sites for restriction enzymes that result in differences in the lengths of the fragments produced by cleavage with the relevant restriction enzyme. RFLPs are used in genetic mapping to link the genome directly to a conventional genetic marker.

Restriction map: A linear array of sites on DNA cleaved by various restriction enzymes.

Reverse tandem inversions: Two inverted segments are adjacent to each other but mutually interchanged.

Ribosome: Small cellular components composed of specialized ribosomal RNA and protein; site of protein synthesis.

Ring bivalent: A ring-shaped chromosome association at diakinesis or metaphase-I of meiosis with a chiasma in both arms.

Ring chromosome: A physically circular chromosome, produced as a result of chromosomal structural changes and usually unstable mitotically and meiotically.

RNA: Ribonucleic acid; long chain of nucleotides connected by phosphate-to-sugar bonds.

S-phase: Synthesis of DNA occurs in eukaryotic cell cycle.

S^1 nuclease: An enzyme that specifically degrades unpaired (single-stranded) sequences of DNA.

Salivary gland chromosomes: In certain tissues of insects (flies, mosquitoes, and midges) of *Diptera*, polytene chromosomes are found in the interphase nuclei. Many extra replications of each chromosomes occur in a single nucleus (endopolyploidy), and all lined up together in parallel fashion (polytene), resulting in thick chromosomes. Discovered and identified first in *Drosophila melanogaster.*

SAT-chromosome: A chromosome with a secondary constriction, usually associated with a nucleolus.

Satellite: A small-terminal segment of chromosome, mostly associated with the nucleolus and is known as a nucleolus organizer chromosome.

Satellite DNA: This is highly repetitive, mostly found in heterochromatin and is usually not transcribed. Many satellite DNA sequences are localized to the kinetochore.

Secondary chromosomes: A chromosome consists of both arms homologous (an isochromosome).

Secondary constriction: This is usually associated with the regions when the nucleolus is formed or associated (nucleolus organizer).

Secondary trisomic: An individual carrying an extra isochromosome in addition to its normal somatic chromosome complement.

Second division restitution (SDR): Due to premature cytokinesis before the second meiotic division takes place; sister chromatids end up in the same nucleus.

Second division segregation: First meiotic division is equational and second meiotic division is reducational.

Seed: The fertilized and ripened ovule with an embryo and generally with a food reserve (endosperm or cotyledons).

Segmental allopolyploidy: Characterized by genomes intermediate in degree of similarity and generally exhibiting preferential pairing ($B_1B_1B_2B_2$).

Segregation: The migration of homologous chromosomes into daughter nuclei during meiosis resulting in separation of the genes due to recombination in the offspring.

Sequence-tagged sites (STSs): This is a short stretch of genomic sequence that can be detected by the PCR.

Sequencing: Determination of the order of nucleotides (base sequences) in a DNA or RNA molecule or the order of aminoacids in a protein.

Sex chromosomes: The chromosomes that determine the sex of an organism (X and Y chromosomes).

Sex linkage: The inheritance of a sex chromosome is coupled with that of a given gene located on the X-chromosome.

Sexual reproduction: This requires meiosis and fertilization alternating in a cycle.

Simple Sequence DNA: This is equal to satellite DNA.

Simplex: Aaaa.

Single cross: Crossing between two genetically different parents.

Sister chromatids: The copies of a chromosome produced by its replication during interphase.

SMC: Sperm or spore mother-cell.

Somatic cell: The cells other than the germ cells.

Somatic chromosome doubling: Doubling the chromosomes of vegetative tissues of an organism experimentally such as by colchicine treatment.

Somatic crossing-over: Crossing-over during mitosis of somatic cells.

Somatic mutation: A mutation which occurs in the somatic tissues and is not inherited.

Somatic pairing: Synapsis of homologous chromosomes in the somatic cells.

Spatial order: Chromosomes or genomes may occupy certain domains within the nucleus and/or dividing cells.

Spermatogonia: These cells are in testes, divide mitotically and produce group of diploid primary sporocyte. Primary sporocytes divide meiotically generating two secondary spermatocytes and produce four spermatids after second mitotic division.

Spindle: A bipolar-oriented structure made of protein fibers, organized by centrosomes or kinetochores that functions in orientation, co-orientation, and separation of chromosomes during metaphase and anaphase.

Spindle attachment region: Kinetochore.

S period: The period of DNA (replication) synthesis during interphase.

Spore: A reproductive cell in plant, the product of meiosis and produces gametes (microspore, mega-, or macrospore) after three mitoses.

Sporocyte: A spore mother-cell.

Sporogenesis: Formation of haploid spores in the higher plants.

Sporophyte: A part of the life cycle of plants where spores are produced.

Sporulation: Formation of spores.

Stamen: The pollen-bearing organ of the flower, consisting of filament and anther. It is a basic unit of androecium.

Staminate: A flower with stamens; the carpels being rudimentary or suppressed.

Sterility: Failure of plants to produce function gametes or viable zygotes.

Stigma: The region of a gynoecium usually at the apex of the style on which compatible pollen grains germinate.

Structural homozygosity: Homozygosity for chromosome mutations.

Style: A portion of gynoecium between ovary and stigma through which pollen tubes grow.

Submedian: Kinetochore located nearer one end than the other.

Subspecies: A taxonomic subdivision of species; genomically similar and capable of interbreeding.

Substitution line: A line in which one or more chromosomes are replaced by one or more chromosomes of a donor species.

Supernumerary chromosomes: Chromosomes also known as B-chromosomes or accessory chromosomes. They are largely heterochromatic and believed to be genetically inert or contain few coding genes.

Symmetrical karyotype: All chromosomes are of about the same size.

Synapsis: Pairing of homologous chromosomes during meiosis.

Synaptonemal complex: The morphological structure of synapsed chromosomes under the electron microscope.

Syncyte: A multinucleate cell produced by inhibition of cytokinesis in mitosis, migration of nucleus from one cell to another, and fusion of two cells.

Syndesis: Synapsis.

Synergids: The two haploid cells that lie besides the egg cell in a mature embryo sac.

Syngamy: The union of male and female gametes during fertilization to produce zygote—sexual reproduction.

Synizesis: Clumping of chromosomes in one side of nucleus (= synizetic knot).

Syntenic: Gene loci located on the same chromosome.

Synthetic amphidiploid: An artificially produced amphidiploid.

T-bands: (T = terminal bands), terminal bands are observed after treating chromosomes at high temperature (87°C) at pH 6.7, and after Giemsa staining shows telomere of some chromosomes preferential staining.

Tandem duplication: A fragment chromosome is inserted into the partner chromosome.

Tandem repeats: Multiple copies of the same sequence.

Tandem satellites: Separate constrictions in the larger satellites.

Tapetum: Refers to innermost layer of the pollen-sac wall.

Telocentric chromosome: A chromosome having a terminal kinetochore and one complete arm of a normal chromosome.

Telomere: The end of a chromosome arm.

Telophase I (meiosis): Chromosomes reach to their respective poles, nuclear membrane and nucleolus start to develop, and eventually two-daughter nuclei each containing the haploid chromosomes are generated.

Telotrisomic: An individual with a normal chromosome complement plus an extra telocentric chromosome is called telotrisomic.

Terminal chromosomal association: The end to end chromosome pairing.

Terminalization: The movement of chiasma toward the ends of the chromosomes.

Tertiary chromosome: A chromosome formed by interchange between nonhomologous chromosomes.

Tertiary gene pool: Gene transfer from species of tertiary gene pool to the cultivated crop of primary gene pool usually requires special crossing and embryo rescue techniques in order to obtain viable derived lines.

Tertiary trisomics: An individual containing an interchanged nonhomologous chromosome in addition to the normal somatic chromosome complement.

Tetrad (meiosis): Four uninuleate cells with the haploid chromosome number produced at the end of meiosis.

Tetraploid: An organism having four basic chromosome sets ($4x$) in the nucleus.

Tetrasome: An organism with one chromosome in the complement represented four times.

Tetraspory: Meiotic karyokinesis occurs but cytokinesis does not.

Tissue culture: The maintenance or growth of tissues *in vitro* in a way that may allow further differentiation and preservation of architecture, or function, or both.

Transgenic: A plant or animal modified by genetic engineering to contain DNA from an external source is called transgenic.

Translocation: A chromosomal structural aberration involving the reciprocal exchange of terminal segments of nonhomologous chromosomes.

Translocation tester set: A series of more or less defined homozygous reciprocal translocations.

Transposable elements: Gene loci capable of being transposed from one spot to another within and among the chromosomes of the complement.

Transposition: The transfer of a chromosome segment to another position due to intra- and interchromosomal structural changes.

Triploid: An organism with three basic sets ($3x$) of chromosomes in the nucleus may be auto- or alloploid.

Trisomic: An organism containing a normal chromosome complement and one extra chromosome.

Trisomic analysis: A method to locate a gene on the chromosome by using trisomics; if a gene is associated with the chromosome, the disomic ratio is modified.

Triplex: AAAa.

Trivalent: An association of three homologous chromosomes.

Unbalanced translocation: A type of chromosome translocation in which a loss of chromosomal segments results in a deleterious effect.

Uninemy: Single strandedness of DNA in the chromosome.

Uniparental inheritance: A kind of inheritance in which only one parent provides genes to the progeny; a phenomenon, usually exhibited by extra nuclear genes, in which all progenies have the phenotype of only one parent.

Unipolar: A cell spindle with one pole.

Univalent: An unpaired chromosome during meiosis.

Univalent shift: A meiotic irregularity in monosomics that produces a monosomic plant deficient for a chromosome other than that originally deficient in the monosomic parent.

Unreduced gametes: Gametes not resulting from common meiosis, and so showing the number of chromosomes per cell that is characteristic of a sporophyte; they spontaneously arise as a consequence of irregular division in anaphase-I of meiosis resulting in spontaneous (meiotic) polyploidization.

Valance crosses: Crossing of individuals of different ploidy levels.

Weismannism: A concept proposed by August Weismann that acquired traits are not inherited and only changes in the germplasm are transmitted from generation to generation.

Wide hybridization: Cross combination between taxonomically distantly related species or genera.

Wild type: Normal phenotype of an organism.

x: A symbol for basic number of chromosomes in a polyploid species.

X-chromosome: A chromosome concerned with the determination of sex.

Xenia: Effect of genotype of pollen on embryo and endosperm.

Xenogamy: Cross-pollination

Xerophyte: A type of plant of arid habitats.

Xylem: A specialized vascular tissue that conducts water and nutrients in plants and provides mechanical support in vascular plants.

Yeast: Yeast is included in fungi that produce asci and ascospores (ascomycetes). Yeast may be haploid or diploid, unicellular organism, and multiplies by budding (*Saccharomyces cerevisiae*).

Y-chromosome: The sex chromosome found only in males in heterozygous condition.

Zygonema (meiosis): The stage at which homologues begin to pair.

Zygote: A fertilized egg.

Zygotic embryo: An embryo derived from fusion of male and female gametes.

Zygotic or somatic chromosome number: $2n$.

[a] Allard, 1960; Chapman and Peat 1992; Darlington and Mather 1949; Fahn 1990; Fehr 1987; Lewin 1990; Nelson et al. 1970; Rieger et al. 1976; http://linkage.rockefeller.edu; Schlegel, R.H.J. 2010. *Dictionary of Plant Breeding.* 2nd ed. CRC Press, Boca Raton, FL.

Index

Printed and bound by CPI Group (UK) Ltd, Croydon, CR0 4YY

17/10/2024

01775709-0010